A Concise Introduction to Robot Programming with ROS2

A Concise Introduction to Robot Programming with ROS2 provides the reader with the concepts and tools necessary to bring a robot to life through programming. It will equip the reader with the skills necessary to undertake projects with ROS2, the new version of ROS. It is not necessary to have previous experience with ROS2 as it will describe its concepts, tools, and methodologies from the beginning.

Key Features

- Uses the two programming languages officially supported in ROS2 (C++, mainly, and Python)
- Approaches ROS2 from three different but complementary dimensions: the Community, Computation Graph, and the Workspace
- Includes a complete simulated robot, development and testing strategies, Behavior Trees, and Nav2 description, setup, and use
- A GitHub repository with code to assist readers

It will appeal to motivated engineering students, engineers, and professionals working with robot programming.

Francisco Martín Rico, Doctor Engineer in Robotics, is an Associate Professor at the Rey Juan Carlos University, where he leads the Intelligent Robotics Lab and teaches courses on Software Architectures and Middlewares for Robots, Mobile Robotics, Planning or Cognitive Systems. He is a reputed member of the ROS community, authoring and contributing to reference packages like ROS2 Planning System (PlanSys2) and Nav2. He has recently received the Best ROS Developer 2022 award.

A Concise Introduction to Robot Programming with ROS2

Francisco Martín Rico

CRC Press
Taylor & Francis Group
Boca Raton London New York

CRC Press is an imprint of the
Taylor & Francis Group, an **informa** business
A CHAPMAN & HALL BOOK

First edition published 2023
by CRC Press
6000 Broken Sound Parkway NW, Suite 300, Boca Raton, FL 33487-2742

and by CRC Press
4 Park Square, Milton Park, Abingdon, Oxon, OX14 4RN

CRC Press is an imprint of Taylor & Francis Group, LLC

ISBN: 978-1-032-26720-3 (hbk)
ISBN: 978-1-032-26465-3 (pbk)
ISBN: 978-1-003-28962-3 (ebk)

DOI: 10.1201/9781003289623

Typeset in Minion
by KnowledgeWorks Global Ltd.

Publisher's note: This book has been prepared from camera-ready copy provided by the authors.

Access the Support Material: https://github.com/fmrico/book ros2

Contents

List of Figures

CHAPTER 1

Introduction

ROBOTS must be programmed to be useful. It is useless for a robot to be a mechanical prodigy without providing it with software that processes the information from the sensors to send the correct commands to the actuators to fulfill the mission for which it was created. This chapter introduces the middlewares for programming robots and, in particular, to *ROS2*[8], which will be the one used in this book.

First of all, nobody starts programming a robot from scratch. Robot software is very complex since we have to face the problem that a robot performs tasks in a real, dynamic, and sometimes unpredictable world. It also must deal with a wide variety of models and types of sensors and actuators. Implementing the necessary drivers or adapting to new hardware components is a titanic effort doomed to failure.

Middleware is a layer of software between the operating system and user applications to carry out the programming of applications in some domains. Middleware usually contains more than libraries, including development and monitoring tools, and a development methodology. Figure 1.1 shows a schematic of a system that includes middleware for developing applications for robots.

Robot programming middlewares provide drivers, libraries, and methodologies. They also usually offer development, integration, execution, and monitoring tools. Throughout the history of Robotics, a multitude of robot programming middlewares have emerged. Few of them have survived the robot for which they were designed or have expanded from the laboratories where they were implemented. There are notable examples (Yarp[5], Carmen[6], Player/Stage[2], etc), although without a doubt the most successful in the last decade has been ROS[7], which is currently considered

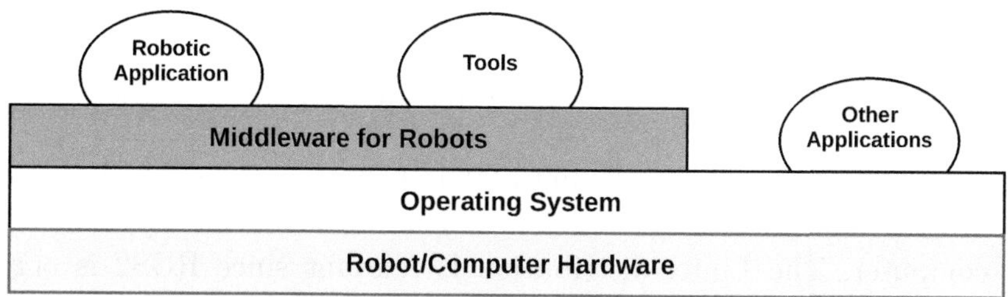

Figure 1.1: Representation of software layers in a robot.

DOI: 10.1201/9781003289623-1

a standard in the world of robot programming. Technically, there are similarities between the different middlewares: most are based on Open Source, many provide communication mechanisms for distributed components, compilation systems, monitoring tools, etc. The big difference is the ROS developers community around the world. There are also leading companies, international organizations, and universities worldwide in this community, providing a vast collection of software, drivers, documentation, or questions already resolved to almost any problem that may arise. Robotics can been defined as "the art of integration", and ROS offers a lot of software to integrate, as well as the tools to do so.

This book will provide the skills necessary to undertake projects in ROS2, the new version of ROS. It is unnecessary to have previous experience in ROS2 since we will describe its concepts, tools, and methodologies from the beginning without the need of previous experience. We will assume average Linux and programming skills. We will use the two programming languages officially supported in ROS2 (C++ and Python), which coincide with the languages most used in general in Robotics.

1.1 ROS2 OVERVIEW

The meaning of the acronym ROS is *Robot Operating System*. It is not an operating system that replaces Linux or Windows but a middleware that increases the system's capabilities to develop Robotic applications. The number *2* indicates that it is the second generation of this middleware. The reader who already knows the first version of ROS (sometimes referred to as ROS1) will find many similar concepts, and there are already several teaching resources[1] for the ROS1 programmer who lands on ROS2. In this book, we will assume no previous knowledge of ROS. It will be more and more common for this to happen, as there are now more and more reasons to learn ROS2 directly instead of going through ROS1 first.

Also, there are already some excellent official ROS2 tutorials, so the approach in this book is intended to be different. The description will be complete and with a methodology oriented to developing robotic applications that make the robot do something "smart", from robotic engineer to robotic engineer, emphasizing essential issues that come from experience in the development of software in robots. It will not hurt for the reader to explore the tutorials available to complete their training and fill in the gaps that do not fit in this book:

- Official ROS2 tutorials: https://docs.ros.org/en/foxy/Tutorials.html
- The Robotics Back-End tutorials: https://roboticsbackend.com/category/ros2/
- ROS2 for ROS2 developers: https://github.com/fmrico/ros_to_ros2_talk_examples

The starting point is a Linux Ubuntu 20.04 LTS system installed on a computer with an AMD64-bit architecture, the most extended one in a personal laptop or desktop computer. The Linux distribution is relevant since ROS2 is organized in *distributions*. A distribution is a collection of libraries, tools, and applications whose

[1] https://github.com/fmrico/ros_to_ros2_talk_examples

versions are verified to work together correctly. Each distribution has a name and is linked to a version of Ubuntu. The software in a distribution is also guaranteed to work correctly with the software version present on the system. It is possible to use another Linux distribution (Ubuntu, Fedora, Red Hat ...), but the reference is Ubuntu. ROS2 also works on Windows and Mac, but this document focuses on Linux development. We will use the ROS2 Foxy Fitzroy version, which corresponds to Ubuntu 20.04.

In this book, we will approach ROS2 from three different but complementary dimensions:

- **The Community**: The ROS community is a fundamental element when developing applications for robots with this middleware. In addition to providing technical documentation, there is a vast community of developers who contribute with their own applications and utilities through public repositories, to which other developers can contribute. Another member of the community may have already developed something you need.
- **Computation Graph**: The Computational Graph is a running ROS2 application. This graph is made up of nodes and arcs. The *Node*, the primary computing units in ROS2, can collaborate with other nodes using several different communication paradigms to compose a ROS2 application. This dimension also addresses the monitoring tools, which are also nodes that are inserted in this graph.
- **The Workspace**: The Workspace is the set of software installed on the robot or computer, and the programs that the user develops. In contrast to the Computational Graph, which has a dynamic nature, the Workspace is static. This dimension also addresses the development tools to build the elements of the Computational Graph.

1.1.1 The ROS Community

The first dimension of ROS2 to consider is the ROS Community. The Open Source Robotics Foundation[2] greatly enhanced the community of users and developers. ROS2 is not only a robot programming middleware, but it is also a development methodology, a established software delivery mechanisms, and a set of resources made available to members of the ROS community.

ROS2 is fundamentally *Open Source*, which means that it is software released under a license in which the user has rights of use, study, change, and redistribution. Many Open Source licenses modulate certain freedoms on this software, but essentially we can assume these rights. The most common licenses for ROS2 software packages are *Apache 2* and *BSD*, although developers are free to use others.

ROS2 organizes the software following a federal model, providing the technical mechanisms that make it possible. Each developer, company, or institution can develop their software freely, responsible for managing it. It is also widespread that small

[2]https://www.openrobotics.org

projects create a community around it, and this community can organize decision-making on releasing issues. These entities create software *packages* for ROS2 that they can make available in public repositories or be part of a ROS2 distribution as binaries. Nobody can force these entities to migrate their software to new versions of ROS2. However, the inertia of many essential and popular packages is enough to guarantee their continuity.

The importance of this development modeling is that it fosters the growth of the ROS community. From a practical point of view, this is key to the success of a robot programming middleware. One of the desirable characteristics of this type of middleware is its support for many sensors and actuators. Nowadays, many manufacturers of these components officially support their drivers for ROS2 since they know that there are many potential customers and that there are many developers who check if a specific component is supported in ROS2 before buying them. In addition, these companies usually develop this software in open repositories where users communities can be created reporting bugs and even sending their patches. If you want your library or tool for robots to be widely used, supporting ROS2 may be the way.

The packages in ROS2 are organized in distributions. A ROS2 distribution is made up of a multitude of packages that can work well together. Usually, this implies that it is tied to a specific version of a base system. ROS2 uses Ubuntu Linux versions as reference. This guarantees stability since otherwise, the dependencies of versions of different packages and libraries would make ROS2 a real mess. When an entity releases specific software, it does so for a given distribution. It is common to maintain multiple development branches for each distribution.

ROS2 has released a total of seven distributions to date (January'22), which we can see in Figure 1.2. Each distribution has a name whose initial increases and a different logo (and a different T-shirt model!). An eighth distribution, which is a bit special, called Rolling Ridley, serves as a staging area for future stable distributions of ROS2 and as a collection of the most recent development releases.

Distro name	**Release Data**	**EOL date**	**Ubuntu version**
Galactic Geochelone	May 23rd, 2021	November 2022	Ubuntu 20.04
Foxy Fitzroy	June 5th, 2020	May 2023 (LTS)	
Eloquent Elusor	November 22nd, 2019	November 2020	Ubuntu 18.04
Dashing Diademata	May 31st, 2019	May 2021 (LTS)	
Crystal Clemmys	December 14th, 2018	December 2019	
Bouncy Bolson	July 2nd, 2018	July 2019	Ubuntu 16.04
Ardent Apalone	December 8th, 2017	December 2018	

If you want to contribute your software to a distribution, you should visit the rosdistro repository (https://github.com/ros/rosdistro), and a couple of useful links:

- Contributing: https://github.com/ros/rosdistro/blob/master/CONTRIBUTING.md
- Releasing your package: https://docs.ros.org/en/rolling/How-To-Guides/Releasing/Releasing-a-Package.html

Figure 1.2: ROS2 distributions delivered until Aug 2021.

The Open Source Robotics Foundation makes many resources available to the community, among which we highlight:

- ROS official page. http://ros.org
- ROS2 Documentation Page: https://docs.ros.org/. Each distro has its documentation. For example, at https://docs.ros.org/en/foxy/ you can find installation guides, tutorials, and guides, among others.
- ROS Answers (https://answers.ros.org/). A place to ask questions and problems with ROS.
- ROS Discourse (https://discourse.ros.org/). It is a discussion forum for the ROS community, where you can keep up to date with the community, view release announcements, or discuss design issues. They also have ROS2 user groups in multiple languages.

1.1.2 The Computation Graph

In this second dimension, we will analyze what a robot's software looks like during its execution. This vision will give us an idea of the goal, and we will be able to understand better the why of many of the contents that will follow. This dimension is what we call *Computation Graph.*

A Computation Graph contains ROS2 nodes that communicate with each other so that the robot can carry out some tasks. The logic of the application is in the nodes, as the primary elements of execution in ROS2.

ROS2 makes intensive use of Object-Oriented Programming. A node is an *object* of class `Node`, in general, whether it is written in C++ or Python.

A node can access the Computation Graph and provides mechanisms to communicate with other nodes through 3 types of paradigms:

- **Publication/Subscription**: It is an asynchronous communication, where N nodes publish messages to a topic that reaches its M subscribers. A topic is like a communication channel that accepts messages of a unique type. This type of communication is the most common in ROS2. A very representative case is the node that contains the driver of a camera that publishes images in a topic. All the nodes in a system needing images from the camera to carry out their function subscribe to this topic.

- **Services**: It is a synchronous communication in which a node requests another node and waits for the response. This communication usually requires an immediate response so as not to affect the control cycle of the node that calls the service. An example could be the request to the mapping service to reset a map, with a response indicating if the call succeeded.

- **Actions**: These are asynchronous communications in which a node makes a request to another node. These requests usually take time to complete, and the calling node may periodically receive feedback or the notification that it has finished successfully or with some error. A navigation request is an example of this type of communication. This request is possibly time-consuming, whereby the node requesting the robot to navigate should not be blocked while completing.

The function of a node in a computational graph is to perform processing or control. Therefore, they are considered active elements with some alternatives in terms of their execution model:

- **Iterative execution**: It is popular in the control software for a node to execute its control cycle at a specific frequency. This approach allows controlling how many computational resources a node requires, and the output flow remains constant. For example, a node calculating motion commands to actuators at 20 Hz based on their status.

- **Event-oriented execution**: The execution of these nodes is determined by the frequency at which certain events occur, in this case, the arrival of messages at this node. For example, a node that, for each image it receives, performs detection on it and produces an output. The frequency at which an output occurs depends on the frequency at which images arrive. If no images reach it, it produces no output.

Next, we will show several examples of computational graphs. The legend in Figure 1.3 shows the elements used in these examples.

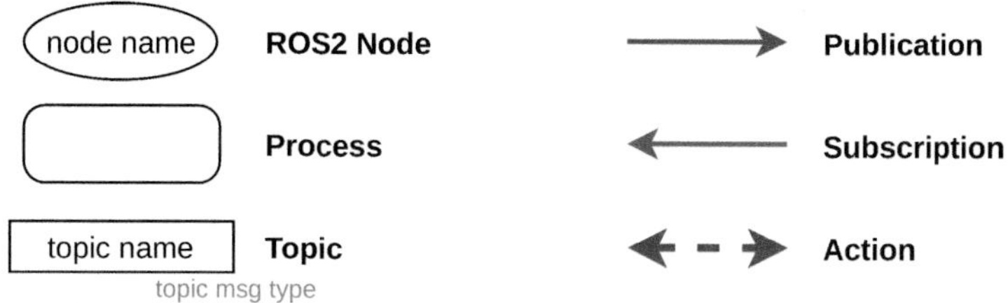

Figure 1.3: Description of symbols used in computer graph diagrams.

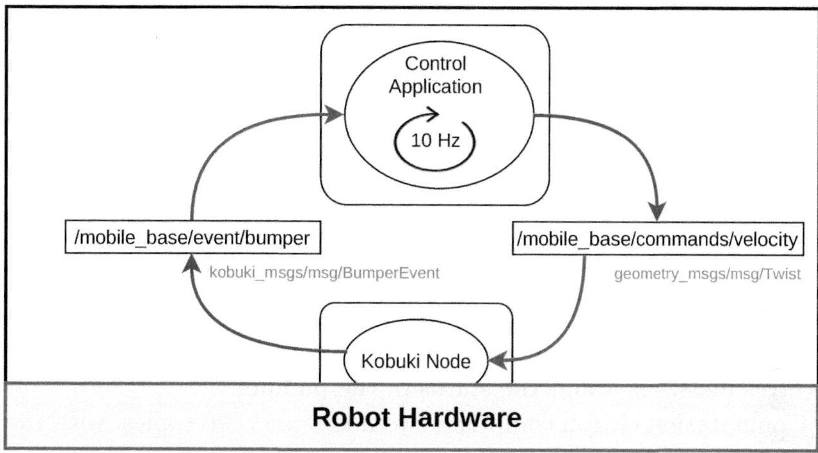

Figure 1.4: Computing graph of a simple control for the Kobuki robot. The control application publishes speeds computed from the information of the bumper to which it subscribes.

The first Computation Graph, shown in Figure 1.4, is a simple example of a program that interacts with a Kobuki[3] robot.

The Kobuki robot driver is a node that communicates with the robot's hardware, probably using a native driver. Its functionality is exposed to the user through various topics. In this case, we have shown only two topics:

- `/mobile_base/event/bumper`: It is a topic in which the kobuki driver publishes a `kobuki_msgs/msg/BumperEvent` message every time one of the bumpers changes state (whether or not it is pressed). All nodes of the system interested on detecting collision with this sensor subscribe to this topic.
- `/mobile_base/commands/velocity`: It is a topic to which the kobuki driver subscribes to adjust its speed. If it does not receive any command in a second, it stops. This topic is of type `geometry_msgs/msg/Twist`. Virtually all robots in ROS2 receive these types of messages to control their speed.

[3]http://kobuki.yujinrobot.com/about2

Deeping into: Names in ROS2

The names of the resources in ROS2 follow a convention very similar to the filesystem in Unix. When creating a resource, for example, a publisher, we can specify its name relative, absolute (begins with "/"), or privately (begins with "~"). Furthermore, we can define a namespace whose objective is to isolate resources from other namespaces by adding the workspace's name as the first component of the name. Namespaces are helpful, for example, in multirobot applications. Let's see an example of the resulting name of a topic based on the node name and the namespace:

name	Result: (node: my_node / ns: none)	Result: (node: my_node / ns: my_ns)
my_topic	/my_topic	/my_ns/my_topic
/my_topic	/my_topic	/my_topic
~my_topic	/my_node/my_topic	/my_ns/my_node/my_topic

Further readings:

- http://wiki.ros.org/Names
- https://design.ros2.org/articles/topic_and_service_names.html

This node runs inside a separate process. The Computation Graph shows another process that subscribes to the bumper's topic, and based on the information it receives, it publishes the speed at which the robot should move. We have set the node's execution frequency to indicate that it makes a control decision at 10 Hz, whether or not it receives messages about the status of the bumper.

This Computation Graph comprises two nodes and two topics, with their respective publication/subscription connections.

Let's evolve the robot and the application. Let's add a laser and a 3D camera (also called RGBD camera). For each sensor, a node must access the sensor and present it with a ROS2 interface. As we said earlier, publishing the data from a sensor is the most convenient way to make this data available in a computational graph.

The application now makes the robot move towards people or objects detected from the 3D image of an RGBD camera. A laser sensor avoids colliding as we move. The Computation Graph shown in Figure 1.5 summarizes the application:

- The control node runs at 20 Hz sending control commands to the robot base. It subscribes to the topic `/scan` to check the obstacles around it.
- The process contains two nodes that detect people and objects, respectively. Both need the image and depth information from the camera to determine the position of detected objects. Each detection is published in two different topics, using a standard message designed for 3D detection.
- The control node subscribes to these topics to carry out its task.

In the last example, the robot used is Tiago[4]. Let's assume that there is only one node that provides its functionality. We use in this example two subscribers (speed commands to move its base and trajectory commands to move its neck) and two publishers (laser information and the 3D image from a RGBD camera).

[4]https://pal-robotics.com/es/robots/tiago/

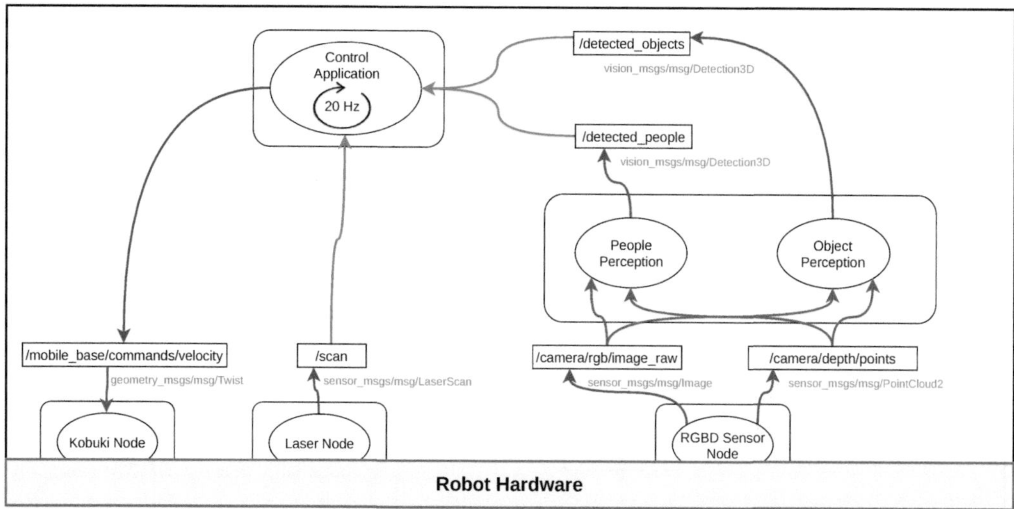

Figure 1.5: Computing graph of a control application that uses the laser data and preprocessed information (people and objects) obtained from the robot's RGBD camera.

The application (Figure 1.6) is divided into two subsystems, each one in a different process that contains the nodes of each subsystem (we have omitted the details of the topics of each subsystem):

- **Behavior subsystem**: It comprises two nodes that collaborate to generate the robot's behavior. There is behavior coordinator (`Coordinator`) and a node that implements an active vision system (`HeadController`). `Coordinator` determines where to look at and which points the robot should visit on a map.
- **Navigation Subsystem**: This example of a navigation subsystem consists of several nodes. The navigation manager coordinates the planner (in charge of creating routes from the robot's position to the destination) and the controller (which makes the robot follow the created route). The planner needs the map provided by a node that loads the environment map and the robot's position that calculates a location node.
- Communication between both subsystems is done using ROS2 actions. The Navigation Behavior sets a goal and is notified when it is complete. It also periodically receives progress toward the destination. Actions are also used to coordinate the planner and controller within the navigation system.

Throughout this subsection, we have shown computation graphs. Every time we implement an application in ROS2, we design a computational graph. We establish which nodes we need and what their interactions are. We must decide if a node is executed at a specific frequency or if some event causes its execution (request or message). We can develop all the nodes, or include in the Computation Graph nodes developed by third parties.

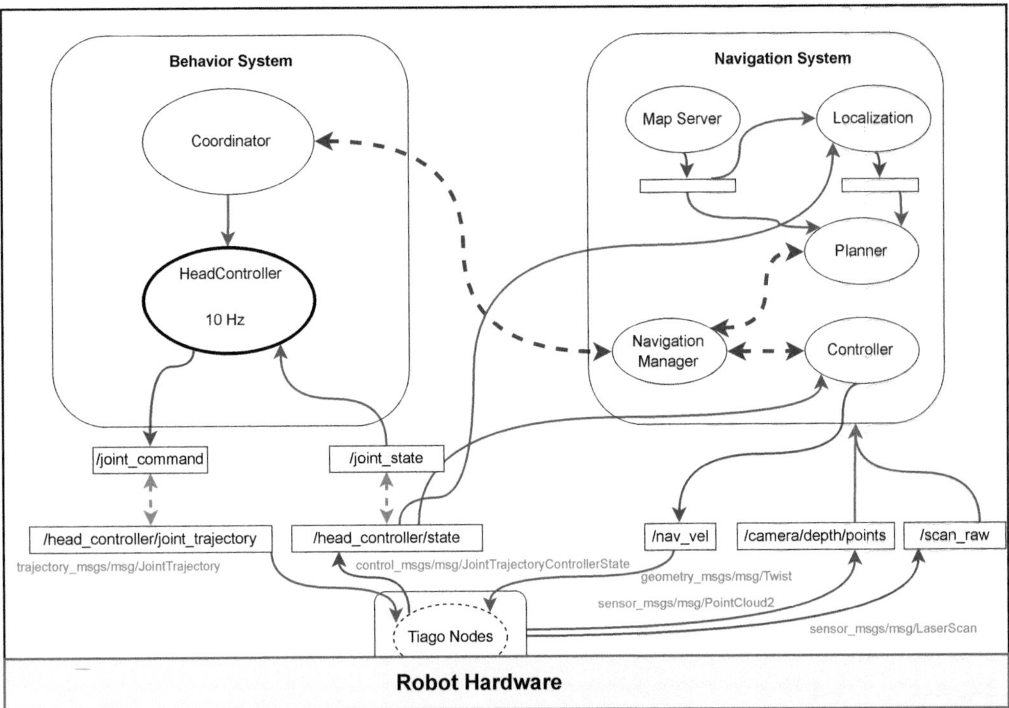

Figure 1.6: Computing graph of behavior-based application for the Tiago robot that uses a navigation subsystem.

Although we can define new message types in our application, ROS2 has defined a set of standard message types that facilitate the interaction of nodes from different developers. It does not make sense, for example, to define a new type of message for images, since there are a lot of third-party software, processing and monitoring tools that consume and produce the type of message considered standard for images, (`sensor_msgs/msg/Image`). Always use existing standard messages whenever possible.

1.1.3 The Workspace

The Workspace dimension approaches ROS2 software from a static point of view. It refers to where the ROS2 software is installed, organized, and all the tools and processes that allow us to launch a computing graph. This includes the build system and node startup tools.

The fundamental element in this dimension is the *package*. A package contains executables, libraries, or message definitions with a common purpose. Usually, a package *depends* on other packages to run or be built.

Another element of this dimension is the *workspace* itself. A workspace is a directory that contains packages. This workspace has to be activated so that what it contains is available to use.

There can be several workspaces active at the same time. This activation process is accumulative. We can activate an initial workspace, that we call *underlay*. Later, we can activate another workspace that we will call *overlay* because it overlays the previous *underlay* workspace. The overlay package dependencies should be satisfied in the underlay. If a package in the overlay already exists in the underlay, the overlay package hides the one in the underlay.

Usually, the workspace containing the basic ROS2 installation is activated initially. This is the most common *underlay* in a ROS2 system. Then, the workspace, where the user is developing their own packages, is activated.

Packages can be installed from sources or with the system installation system. On Ubuntu Linux 20.04, which is the reference in this book, it is carried out with deb packages using tools like *apt*. Each ROS2 package is packaged in a deb package. The names of deb packages in a distribution are easily identifiable because their names start with `ros-<distro>-<ros2 package name>`. In order to access these packages, configure the APT ROS2 repository:

```
$ sudo apt update && sudo apt install curl gnupg2 lsb-release

$ sudo curl -sSL https://raw.githubusercontent.com/ros/rosdistro/master/ros.key
-o /usr/share/keyrings/ros-archive-keyring.gpg

$ echo "deb [arch=$(dpkg --print-architecture)
signed-by=/usr/share/keyrings/ros-archive-keyring.gpg]
http://packages.ros.org/ros2/ubuntu $(source /etc/os-release && echo
$UBUNTU_CODENAME) main" | sudo tee /etc/apt/sources.list.d/ros2.list > /dev/null


$ sudo apt-get update
```

Of course, the installation dependencies of the Deb packages are those of the ROS2 package. The following command shows the ROS2 packages available to install:

```
$ apt-cache search ros-foxy
```

ROS2 Foxy installation. Instructions are located at `https://docs.ros.org/en/foxy/Installation/Ubuntu-Install-Debians.html`. Basically, ROS2 Foxy is installed just typing:

```
$ sudo apt update

$ sudo apt install ros-foxy-desktop
```

All the ROS2 software installed by apt is in `/opt/ros/foxy`. On an Ubuntu 20.04 system, installing the ROS2 Galactic version or the ROS Noetic Ninjemys version is also possible. If both are installed, they are in `/opt/ros/galactic` and `/opt/ros/noetic`, respectively. We could even install one of these ROS distributions by compiling its source code in some other location. Because of this, and because it is not recommended (unless you know what you are doing) to mix ROS distributions, installing a distribution does not "activate" it. The activation is done by executing in a terminal:

```
$ source /opt/ros/foxy/setup.bash
```

This command "activates" the software in `/opt/ros/foxy`. It is common to include this line in `$HOME/.bashrc` so that it is activated by default when opening a terminal:

```
$ echo "source /opt/ros/foxy/setup.bash" >>  /.bashrc
```

It is also convenient to install and configure the *rosdep*[5] tool. This tool discovers dependencies not satisfied in a set of source packages and installs them as deb packages. We only need to run these commands once after installation:

```
$ sudo rosdep init

$ rosdep update
```

Typically, the user creates a directory in his $HOME directory that contains the sources of the packages he is developing. Let's create a workspace only by creating a directory with a `src` directory within. Then, add the example packages that we will use throughout this book.

```
$ cd

$ mkdir -p bookros2_ws/src

$ cd bookros2_ws/src

$ git clone-b foxy-devel https://github.com/fmrico/book_ros2.git
```

If we explore the content that we have added under `src`, we will be able to see a collection of directories. Packages are those that have a `package.xml` file in their root directory.

In this workspace, there are many packages with dependencies on other packages not part of the ROS2 Foxy distribution. To add the sources of these packages to the workspace, we will use the *vcstool*[6]:

```
$ cd ~/bookros2_ws/src

$ vcs import . < book_ros2/third_parties.repos
```

The command `vcs` reads a list of repositories from a `.repos` file and clones them into the specified directory. Before building, let's use `rosdep` to install any package missing to build the entire workspace:

```
$ cd ~/bookros2_ws

$ rosdep install --from-paths src --ignore-src -r -y
```

Once the sources of the example packages with their dependencies are in the working workspace, build the workspace, always from its root, using the `colcon`[7] command:

[5] http://wiki.ros.org/rosdep

[6] https://github.com/dirk-thomas/vcstool

[7] https://colcon.readthedocs.io/en/released/index.html

```
$ cd ~/bookros2_ws

$ colcon build --symlink-install
```

Check that three directories have been created in the worspace root:

- **build**: It contains the intermediate files of the compilation, as well as the tests, and temporary files.
- **install**: It contains the compilation results, along with all the files necessary to execute them (specific configuration files, node startup scripts, maps ...). Building the workspace using the `--symlink-install` option, creates a symlink to their original locations (in `src` or `build`), instead of copying. This way, we save space and can modify certain configuration files directly in `src`.
- **log**: Contains a log of the compilation or testing process.

Deeping into: `colcon`

`colcon` (collective construction) is a command line tool for building, testing, and using multiple software packages. With colcon, you can compile ROS1, ROS2 and even plain cmake packages. It automates the process of building and set up the environment to use the packages.
Further readings:

- https://design.ros2.org/articles/build_tool.html
- https://colcon.readthedocs.io/
- https://vimeopro.com/osrfoundation/roscon-2019/video/379127725

To clean/reset a workspace, simply delete these three directories. A new compilation will regenerate them.

In order to use the content of the workspace, activate it as an overlay, in a similar way to how the underlay was activated:

```
$ source ~/bookros2_ws/install/setup.bash
```

It is common to include this line, as well as the underlay, in `$HOME/.bashrc` so that it is activated by default when opening a terminal:

```
$ echo "source ~/bookros2_ws/install/setup.bash" >>  /.bashrc
```

1.2 THE ROS2 DESIGN

Figure 1.7 shows the layers that compose the design of ROS2. The layer immediately below the nodes developed by the users provides the programmer with the API to interact with ROS2. Packages which nodes and programs are implemented in C++ use the C++ client libraries, *rclcpp*. Packages in python use *rclpy*.

Rclcpp and rclpy are not completely independent ROS2 implementations. If so, a node in Python could have different behavior than one written in C++. Both rclcpp and rclpy use rcl, which implements the basic functionality of all ROS2 elements.

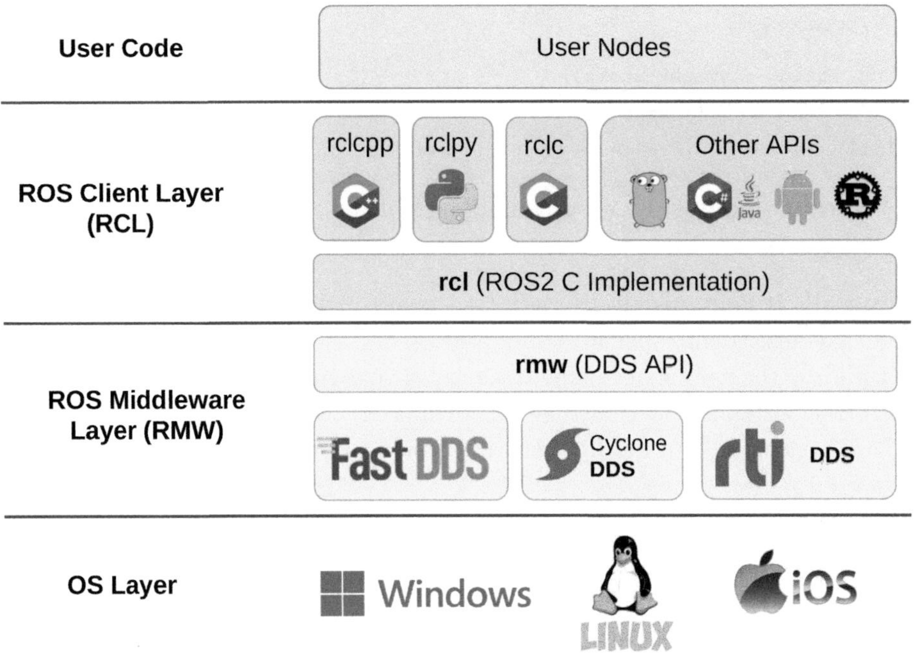

Figure 1.7: ROS2 layered design.

Rclcpp and rclpy adapt this functionality to the particularities of each language, along with certain things that must be done at that level, such as the threads model.

Any client library for another language (Rust[8], Go[9], Java[10], .NET[11], ...), should be built on top of rcl.

Rcl is the core of ROS2. No one uses it directly for their programs. There is a C client library for ROS2 called rclc if the user want to develop C nodes. Although it is written in the same language as rcl, it still has to complete some functionality and make ROS2 programming less arid than programming using rcl directly.

A crucial component of ROS2 is communications. ROS2 is a distributed system whose computing graph has nodes that can be spread over several machines. Even with all the software running on a single robot, nodes are running on the operator's PC to monitor and control the robot's operation.

ROS2 has chosen Data Distribution Service (DDS)[12] for its communications layer, a next-generation communications middleware implemented over UDP. It allows the exchange of information between processes with real-time characteristics, security capabilities, and custom quality of service of each connection. DDS provides a publication/subscription communications paradigm, providing a mechanism to discover

[8] https://github.com/ros2-rust/ros2_rust

[9] https://github.com/tiiuae/rclgo

[10] https://github.com/ros2-java/ros2_java

[11] https://github.com/ros2-dotnet/ros2_dotnet

[12] https://www.omg.org/omg-dds-portal/

publishers and subscribers without needing a centralized service automatically. This discovery is made using multicast, although subsequent connections are unicast.

There are several DDS vendors, including FastDDS[13], CycloneDDS[14], or RTI[15] Connext. All of them fully or partially implement the DDS standard defined by the OMG[16]. ROS2 can use all of these DDS implementations. Very few ROS2 developers could notice when using one or the other. Still, when we begin to require high performance in latency, amount of data, or resources used, we can find some differences that make us choose one with objective criteria.

The APIs of these DDS implementations do not have to be the same. In fact, they are not. For this reason, and to simplify the rcl layer, an underlying layer called rmw has been implemented, which presents the rcl programmer with a unified API to access the functionality of each supported DDS implementation. Selecting which DDS to use is trivial, requiring just modifying an environment variable, `RMW_IMPLEMENTATION`.

The official version of DDS in the Foxy distribution is FastDDS, while in Galactic, it is CycloneDDS. These vendors compete (in what ironically came to be called the *DDS Wars*) to be the predominant implementation. Hopefully, the main beneficiary of this competition will be the ROS2 community.

1.3 ABOUT THIS BOOK

This book is intended to be a journey through programming robots in ROS2, presenting several projects where the main ROS2 concepts are applied. Prior knowledge of ROS/ROS2 is not needed. Many of the concepts we will see will sound very familiar to ROS1 programmers. They will find interesting the changes that ROS2 presents concerning the previous version.

We will use C++ as a vehicular language, although we will include one in Python in our first examples. We can develop complex and powerful projects in Python but in my experience with robots, I prefer to use a compiled language rather than an interpreted one. Similarly, the concepts explained with C++ are equally valid with Python. Another decision is to use Linux (specifically Ubuntu GNU / Linux 20.04 LTS) instead of Windows or Mac since it is the reference platform in ROS2 and the one that I consider predominant in Robotics.

I will assume that the reader is a motivated engineering student or an experienced engineer/professional. We will be using many C++ features up to C++17, including smart pointers (`shared_ptr` and `unique_ptr`), containers (vector, list, map), generic programming, and more. I will try to explain complex code parts from a language point of view, but the less advanced reader may need to consult some references[17,18]. I also count on the reader to know CMake, Git, gdb, and other common tools developers use in Linux environments. It can be a great time to learn it if you

[13]https://github.com/eProsima/Fast-DDS
[14]https://github.com/eclipse-cyclonedds/cyclonedds
[15]https://www.rti.com/products/dds-standard
[16]https://www.omg.org/
[17]https://en.cppreference.com/w/
[18]https://www.cplusplus.com/

do not know it because everything used in this book is what a robot programmer is expected to know.

This book is mainly read sequentially. It would be difficult for a beginner in ROS2 to follow the concepts if chapters were skipped. At some points, I will include a text box like this:

Deeping into: Some topic

Some explanation.

This box indicates that in the first reading, it can be skipped and returned later to deepen in some concepts.

Throughout the book, I will type shell commands. ROS2 is used from the shell mainly, and it is important that the user master these commands. I will use these text boxes for commands in the terminal:

```
$ ls /usr/lib
```

This book is not intended to be a new ROS2 tutorial. The ones on the official website are great! In fact, there are many concepts (services and actions) that are best learned in these tutorials. This book wants to teach ROS2 by applying concepts to examples in which a simulated robot performs some mission. Also, we want to teach not only ROS2 but also some general concepts in Robotics and how they are applied in ROS2.

Therefore, in this book, we will analyze much code. I have prepared a repository with all the code that we will use in:

`https://github.com/fmrico/book_ros2`

At the end of each chapter, I will propose exercises or improvements to deepen the subject. If you manage to solve it, it can be uploaded to the official repository of the book, in a separate branch with a description and with your authorship. Do this by making a pull request to the official book repository. If I have time (I hope so), I would be happy to review it and discuss it with you.

Even the existence of this repository, in order to make the book self-contained, I have added the source code of all this software in Annex A. When it comes to indicating what the structure of a package is, I will use this box:

Package `my_package`

```
my_package/
├── CMakeLists.txt
└── src
    └── hello_ros.cpp
```

To show source code I will use this other box:

src/hello_ros.cpp

```
#include <iostream>

int main(int argc, char * argv[]) {
  std::cout << "hello ROS2" << std::endl;

  return 0;
}
```

Moreover, when it's just snippet of code, I will use this kind of box, unnumbered:

```
std::cout << "hello ROS2" << std::endl;
```

I hope you enjoy this book. Let's start our journey along programming robots with ROS2.

CHAPTER 2

First Steps with ROS2

THE previous chapter introduced the fundamental theoretical concepts of ROS2, in addition to installing ROS2. In this chapter, we begin to practice with ROS2 and learn the first ROS2 concepts.

2.1 FIRST STEPS WITH ROS2

ROS2 has been already installed, and it is activated both the underlay (`/opt/ros/foxy`) and the overlay (`~/bookros2_ws`), by adding a source instruction to `~/.bashrc`. Check it typing:

```
$ ros2

usage: ros2 [-h] Call 'ros2 <command> -h' for more detailed usage. ...
ros2 is an extensible command-line tool for ROS 2.

...
```

If the underlay is activated, this command will be found.

ros2 is the main command in ROS2. It allows to interact with the ROS2 system to obtain information or carry out actions.

```
ros2 <command> <verb> [<params>|<option>]*
```

To obtain the list of available packages, type:

```
$ ros2 pkg list

ackermann_msgs
action_msgs
action_tutorials_cpp
...
```

In this case, **pkg** manages ROS2 packages. The **list** verb obtains the list of packages in the underlay or any overlay.

DOI: 10.1201/9781003289623-2

Deeping into: roscli

ros2cli is the ROS2 command line interface tool. It is modular and extensible, so that more functionality can be added by adding new actions. The standard actions currently are:

`action`	`extension_points`	`node`	`test`
`bag`	`extensions`	`param`	`topic`
`component`	`interface`	`pkg`	`wtf`
`launch`	`run`	`daemon`	`lifecycle`
`security`	`doctor`	`multicast`	`service`

Further readings:

- https://github.com/ros2/ros2cli
- https://github.com/ubuntu-robotics/ros2_cheats_sheet/blob/master/cli/cli_cheats_sheet.pdf

The **ros2** command supports tab-key autocompletion. Type `ros2` and then hit the tab key twice to see the possible verbs. The arguments of a verb can also be discovered with the tab key.

It is also possible to obtain information on a specific package. For example, to get the executable programs from the `demo_nodes_cpp` package:

```
$ ros2 pkg executables demo_nodes_cpp

demo_nodes_cpp add_two_ints_client
demo_nodes_cpp add_two_ints_client_async
demo_nodes_cpp add_two_ints_server
demo_nodes_cpp allocator_tutorial
...
demo_nodes_cpp talker
...
```

Execute one of them with the command using the **run** verb, which requires two arguments: the package where the executable is and the name of the executable program: The name of this package indicates that all the programs it contains are written in C++.

```
$ ros2 run demo_nodes_cpp talker

[INFO] [1643218362.316869744] [talker]: Publishing: 'Hello World: 1'
[INFO] [1643218363.316915225] [talker]: Publishing: 'Hello World: 2'
[INFO] [1643218364.316907053] [talker]: Publishing: 'Hello World: 3'
...
```

As can be seen, when specifying the program to be executed with the package name and executable name, it is not necessary to know exactly where the programs are, nor to execute them in any specific location.

If everything went well, "Hello world" messages appear in the terminal with a counter. Keep this command running and open another terminal to see what this executable is doing. It is common in ROS2 to have several terminals open simultaneously, so it is essential to organize them well on the screen to avoid getting lost. The small Computation Graph that has been created is shown in Figure 2.1.

Check the nodes that are currently running using the **node** verb and its **list** argument, executing in another terminal:

Figure 2.1: Computation Graph for the `Talker` node.

```
$ ros2 node list

/talker
```

This command confirms that there is only one node called `/talker`. The names of the resources in ROS2, as is the case of the nodes, have a similar format to the files in a Linux system. The slash separates parts of the name, starting with the `/` root.

The node `/talker` does not just print an information message through the terminal. It is also publishing messages on a topic.

Check, while the node `/talker` is running, what topics are in the system. For this, use the **`topic`** verb with its **`list`** argument.

```
$ ros2 topic list

/chatter
/parameter_events
/rosout
```

There are several topics, including `/chatter`, which is the one that publishes `/talker`. Use the **`info`** parameter of the **`node`** verb to get more information:

```
$ ros2 node info /talker

/talker
   Subscribers:
       /parameter_events: rcl_interfaces/msg/ParameterEvent
   Publishers:
       /chatter: std_msgs/msg/String
       /parameter_events: rcl_interfaces/msg/ParameterEvent
       /rosout: rcl_interfaces/msg/Log
   Service Servers:
...
```

The output shows several publishers, which coincide with the topics shown by the previous command since there are no more nodes in the system.

As we have said, each topic supports messages of only one type. The previous commands already showed the type, although it can be verified by asking the `topic` action directly for the information of a specific topic:

```
$ ros2 topic info /chatter

Type: std_msgs/msg/String
Publisher count: 1
Subscription count: 0
```

Messages are defined in packages that, by convention, end in _msgs. std_msgs/msg/String is the String message defined in the std_msgs package. To check what messages are valid in the system, use the interfaces action and its list argument.

```
$ ros2 interface list

Messages:
    ackermann_msgs/msg/AckermannDrive
    ackermann_msgs/msg/AckermannDriveStamped
    ...
    visualization_msgs/msg/MenuEntry
Services:
    action_msgs/srv/CancelGoal
    ...
    visualization_msgs/srv/GetInteractiveMarkers
Actions:
    action_tutorials_interfaces/action/Fibonacci
    ...
```

The output shows all the types of interfaces through which the nodes can communicate in ROS2. Adding the `-m` option, you can filter only the messages. Note that there are more interfaces than just messages. Services and actions also have a format that we can also inspect with `ros2 interface`.

Check the message format to get the fields contained in the message, and their type:

```
$ ros2 interface show std_msgs/msg/String

... comments
string data
```

This message format has only one field called `data`, of `string` type.

Deeping into: interfaces

A message is made up of fields. Each field has a different type, which can be a basic type (bool, string, float64) or a message type. In this way, It is usual to create more complex messages from simpler messages.

An example is the *stamped* messages. A series of messages, whose name ends in **Stamped**, add a header to an existing message. Check the difference between these two messages:

`geometry_msgs/msg/Point`
`geometry_msgs/msg/PointStamped`

Further readings:

- https://docs.ros.org/en/foxy/Concepts/About-ROS-Interfaces.html

Check the messages currently being published (`/talker` should be still running in the other terminal) in the topic just typing:

```
$ ros2 topic echo /chatter
data: 'Hello World: 1578'
---
data: 'Hello World: 1579'
...
```

Next, execute a program that contains a node that subscribes to the topic `/chatter` and displays the messages it receives on the screen. To execute it, without stopping the program that contains the `/talker` node, we run the `/listener` node, which is in the homonymous program. Although there is a `listener` node in the `demo_nodes_cpp` package, for variety, run the listener from a package where the nodes are implemented in Python:

```
$ ros2 run demo_nodes_py listener
[INFO] [1643220136.232617223] [listener]: I heard: [Hello World: 1670]
[INFO] [1643220137.197551366] [listener]: I heard: [Hello World: 1671]
[INFO] [1643220138.198640098] [listener]: I heard: [Hello World: 1672]
...
```

Now the Computation Graph is made up of two nodes that communicate through the topic `/chatter`. The Computation Graph would look like as shown in Figure 2.2.

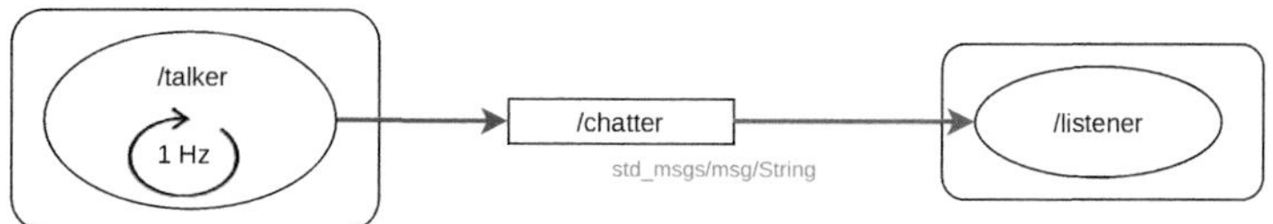

Figure 2.2: Computation Graph for the `Listener` node.

It is also possible to visualize the Computation Graph by running the `rqt_graph tool` (Figure 2.3), which is in the `rqt_graph` package:

```
$ ros2 run rqt_graph rqt_graph
```

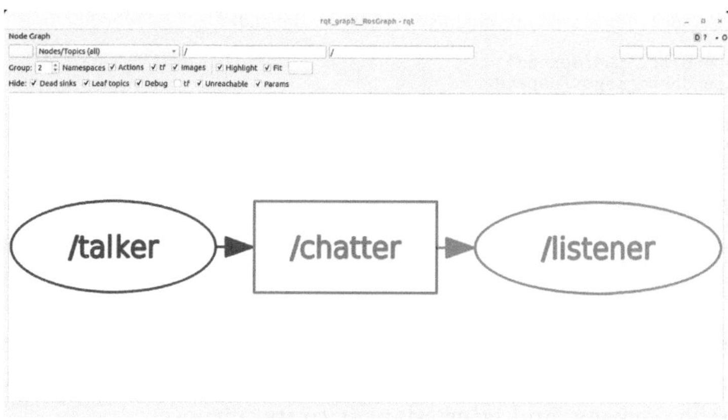

Figure 2.3: Program `rqt_graph`.

Now stop all the programs simply by pressing Ctrl+C in the terminals where they are running.

2.2 DEVELOPING THE FIRST NODE

Up to this point, we have only used software from the packages that are part of the ROS2 base installation. In this section, we will create a package to develop the first node.

The new package will be created in the overlay (`cd ~/bookros2_ws`) to practice creating packages from scratch.

All packages must be in the `src` directory. This time, we use the `ros2` command and the `pkg` verb with the **`create`** option. In ROS2 packages, it is necessary to declare what other packages they depend on, either on this workspace or another, so that the compilation tool knows the order they have to be built. Go to the `src` directory and run:

```
$ cd ~/bookros2_ws/src

$ ros2 pkg create my_package --dependencies rclcpp std_msgs
```

This command creates the skeleton of the basics package, with some empty directories to host the source files of our programs and libraries. ROS2 recognizes that a directory contains a package because it has an XML file called `package.xml`. The **`--dependencies`** option allows you to add the dependencies of this package. For now, we will use rclcpp, which are the C++ client libraries.

package.xml

```xml
<?xml version="1.0"?>
<?xml-model href="http://download.ros.org/schema/package_format3.xsd"
 schematypens="http://www.w3.org/2001/XMLSchema"?>
<package format="3">
  <name>my_package</name>
  <version>0.0.0</version>
  <description>TODO: Package description</description>
  <maintainer email="john.doe@evilrobot.com">johndoe</maintainer>
  <license>TODO: License declaration</license>

  <buildtool_depend>ament_cmake</buildtool_depend>

  <depend>rclcpp</depend>
  <depend>std_msgs</depend>

  <test_depend>ament_lint_auto</test_depend>
  <test_depend>ament_lint_common</test_depend>

  <export>
    <build_type>ament_cmake</build_type>
  </export>
</package>
```

Although `ros2 pkg create` is a good starting point for creating a new package, in practice, it is usually made by duplicating an existing package, immediately changing the name of the package, and later adapt it to its purpose.

As the example is a C++ package, since we have indicated that it depends on rclcpp, in its root, a `CMakeLists.txt` file has also been created that establishes the rules to compile it. We will analyze its content as soon as we add something to compile.

First, Create the program in ROS2, as simple as possible, and call it `src/simple.cpp`. Next boxes contains the package structure, and the source code of `src/simple.cpp`:

Package `my_package`

```
my_package/
├── CMakeLists.txt
├── include
│   └── my_package
├── package.xml
└── src
    └── simple.cpp
```

`src/simple.cpp`

```
1  #include "rclcpp/rclcpp.hpp"
2
3  int main(int argc, char * argv[]) {
4    rclcpp::init(argc, argv);
5
6    auto node = rclcpp::Node::make_shared("simple_node");
7
8    rclcpp::spin(node);
9
10   rclcpp::shutdown();
11
12   return 0;
13 }
```

- `#include "rclcpp/rclcpp.hpp"` allows access to most of the ROS2 types and functions in C++.
- `rclcpp::init(argc, argv)` extracts from the arguments with which this process was launched any option that should be taken into account by ROS2.
- Line 6 creates a ROS2 node. **`node`** is a `std::shared_ptr` to a ROS2 node whose name is `simple_node`.

  The `rclcpp::Node` class is equipped with many aliases and static functions to simplify the code. `SharedPtr` is an alias for `std::shared_ptr<rclcpp::Node>`, and `make_shared` is a static method for `std::make_shared<rclcpp::Node>`.

  The following lines are equivalent, going from a pure C++ statement to one that takes advantage of ROS2 facilities:

```
1. std::shared_ptr<rclcpp::Node> node = std::shared_ptr<rclcpp::Node>(
   new rclcpp::Node("simple_node"));

2. std::shared_ptr<rclcpp::Node> node = std::make_shared<rclcpp::Node>(
   "simple_node");

3. rclcpp::Node::SharedPtr node = std::make_shared<rclcpp::Node>(
   "simple_node");

4. auto node = std::make_shared<rclcpp::Node>("simple_node");

5. auto node = rclcpp::Node::make_shared("simple_node");
```

- In this code, `spin` blocks the execution of the program so that it does not terminate immediately. Its important functionality will be explained in the following examples.

- `shutdown` manages the shutdown of a node, prior to the end of the program in the next line.

Examine the `CMakeLists.txt` file, already prepared to compile the program. Some parts that are not relevant now have been removed for clarity:

CMakeLists.txt

```
cmake_minimum_required(VERSION 3.5)
project(basics)

find_package(ament_cmake REQUIRED)
find_package(rclcpp REQUIRED)

set(dependencies
  rclcpp
)

add_executable(simple src/simple.cpp)
ament_target_dependencies(simple ${dependencies})

install(TARGETS
  simple
  ARCHIVE DESTINATION lib
  LIBRARY DESTINATION lib
  RUNTIME DESTINATION lib/${PROJECT_NAME}
)

if(BUILD_TESTING)
  find_package(ament_lint_auto REQUIRED)
  ament_lint_auto_find_test_dependencies()
endif()

ament_export_dependencies(${dependencies})
ament_package()
```

Identify several parts in this file:

- In the first part, the packages needed are specified with `find_package`. Apart of `ament_cmake`, always needed by `colcon`, just `rclcpp` is especified. It is a good habit to create a `dependencies` variable with the packages that this package depends on since we will have to use this list several times.

- For each executable:

Compile it: Do it with `add_executable`, indicating the name of the result and its sources. Also, use `ament_target_dependencies` to make headers and libraries from other packages accessible for the current target. There is no dependencies with extra libraries, so just using `ament_target_dependencies` is fine.

Install it: Indicate where to install the program produced, which generally does not vary. A single `install` instruction will be valid for programs and libraries of the package.

In general, install everything needed to deploy and run the program. *If it is not installed, it does not exist.*

Compile the workspace:

```
cd ~/bookros2_ws

colcon build --symlink-install
```

As we said before, currently, we must re-source the workspace since we have created a new program, so we open a new terminal and execute:

```
$ ros2 run my_package simple
```

And let's see what happens: absolutely nothing (Figure 2.4).

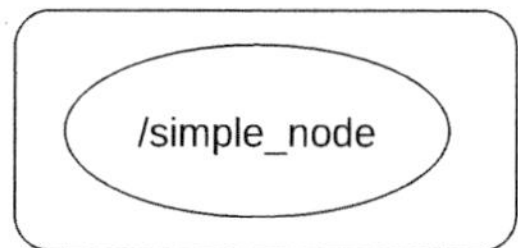

Figure 2.4: Computation Graph for the `Simple` node.

Internally, our program is in the `spin` statement, blocked, waiting for us to finish our program by pressing Ctrl+C. Before doing so, check that the node has been created executing in another terminal:

```
$ ros2 node list

/simple_node
```

Once described how to create a package from scratch. From now on, we will use the packages downloaded from the repository of this book in the previous chapter. This will allow moving faster without getting blocked in small mistakes when building the package, which at this point can prove insurmountable.

2.3 ANALYZING THE BR2_BASICS PACKAGE

Once this process has been seen in detail, continue analyzing the content of the `br2_basics` package, which contains more interesting nodes. The structure of this package is shown in the following box, and the complete source code can be found in the annexes and in the book repository:

Package br2_basics

```
br2_basics
├── CMakeLists.txt
├── config
│   └── params.yaml
├── launch
│   ├── includer_launch.py
│   ├── param_node_v1_launch.py
│   ├── param_node_v2_launch.py
│   ├── pub_sub_v1_launch.py
│   └── pub_sub_v2_launch.py
├── package.xml
└── src
    ├── executors.cpp
    ├── logger_class.cpp
    ├── logger.cpp
    ├── param_reader.cpp
    ├── publisher_class.cpp
    ├── publisher.cpp
    └── subscriber_class.cpp
```

2.3.1 Controlling the Iterative Execution

The previous section described a program containing a node that literally did not do much beyond existing. Program `src/logger.cpp` is more interesting, as it shows some more activity:

src/logger.cpp

```cpp
auto node = rclcpp::Node::make_shared("logger_node");

rclcpp::Rate loop_rate(500ms);
int counter = 0;

while (rclcpp::ok()) {
  RCLCPP_INFO(node->get_logger(), "Hello %d", counter++);

  rclcpp::spin_some(node);
  loop_rate.sleep();
}
```

This code shows the first of the typical strategy to perform a task at a fixed frequency, which is common in any program that performs some control. The control loop is made in a while loop, controlling the rate with an `rclcpp::Rate` object that makes the control loop stop long enough to adapt to the selected rate.

This code uses `spin_some` instead of `spin`, as used so far. Both are to manage the messages that arrive at the node, calling the functions that should handle them. While `spin` blocks waiting for new messages, `spin_some` returns once there are no messages left to handle.

As for the rest of the code, `RCLCPP_INFO` is used, which is a macro that prints information. It's very similar to `printf`, passing as first parameter the node's logger (an object inside nodes to log, got with `get_logger` method). These messages are displayed on the screen and are also published in the topic `/rosout`.

Run this program by typing:

```
$ ros2 run br2_basics logger
[INFO] [1643264508.056814169] [logger_node]: Hello 0
[INFO] [1643264508.556910295] [logger_node]: Hello 1
...
```

The program begins to show messages containing the criticality level of the message, timestamp, the node that produced it, and the message.

As we said before, RCLCPP_INFO also publishes a message of type rcl_interfaces/msg/Log in the topic /rosout, as shown in Figure 2.5. All nodes have a publisher to send the output we generate to this node. It is quite useful when we do not have a console to see these messages.

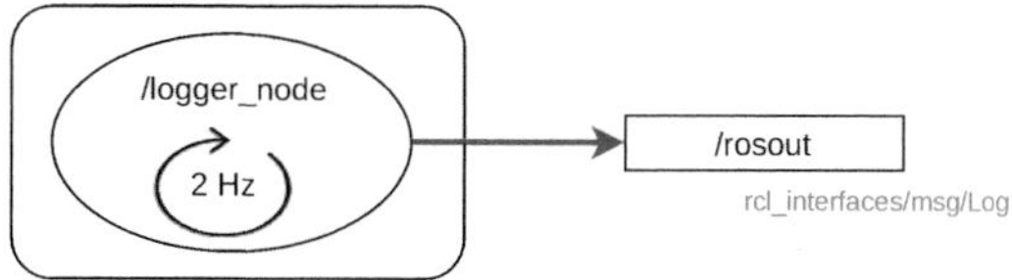

Figure 2.5: Computation Graph for the Logger node.

Take this opportunity to see how to see the messages that are published on a topic:

```
$ ros2 topic echo /rosout
stamp:
    sec: 1643264511
    nanosec: 556908791
level: 20
name: logger_node
msg: Hello 7
file: /home/fmrico/ros/ros2/bookros2_ws/src/book_ros2/br2_basics/src/logger.cpp
function: main
line: 27
---
stamp:
    sec: 1643264512
    nanosec: 57037520
level: 20
...
```

Check the definition of the rcl_interfaces/msg/Log message to verify that the fields shown are the fields of this type of message. In the line field, we have our message:

```
$ ros2 interface show rcl_interfaces/msg/Log
```

Finally, use the rqt_console tool to see the messages that are published in /rosout, as shown in Figure 2.6. This tool, shown in Figure 2.7, is useful when many nodes are generating messages to /rosout, and is useful to filter it by node, by the level of criticality, etc.

```
$ ros2 run rqt_console rqt_console
```

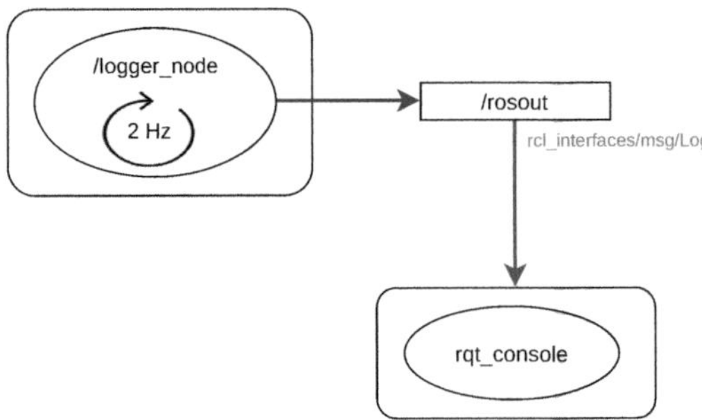

Figure 2.6: `rqt_console` subscribes to `/rosout`, receiving the messages produced by the `Logger` node.

rqt_console__Console - rqt

Console

Load | Save | Pause | Displaying 215 messages | Clear | Fit Columns

#	Message	Severity	Node	Stamp	Location
#215	Hello 245	Info	logger_node	11:57:16.0...	/home/...
#214	Hello 244	Info	logger_node	11:57:15.5...	/home/...
#213	Hello 243	Info	logger_node	11:57:15.0...	/home/...
#212	Hello 242	Info	logger_node	11:57:14.5...	/home/...
#211	Hello 241	Info	logger_node	11:57:14.0...	/home/...
#210	Hello 240	Info	logger_node	11:57:13.5...	/home/...
#209	Hello 239	Info	logger_node	11:57:13.0...	/home/...

Exclude Messages...

Highlight Messages...

...containing: Regex

Figure 2.7: `rqt_console` program.

Test different frequencies by changing the time that object `loop_rate` is created, changing to 100 ms or 1 s, so that the control loop runs at 10 Hz or 1 Hz, respectively.

Do not forget to compile after every change. Use the option `--packages-select` to compile only the package that we have changed, thus saving some time:

```
$ cd ~/bookros2_ws

$ colcon build --symlink-install --packages-select br2_basics
```

From here on, the `cd` command will be omitted. It has become clear that the directory from which all the compilations of a workspace must be carried out is its root.

Deeping into: logging

ROS2 has a logging system that allows generating log messages with increasing severity levels: DEBUG, INFO, WARN, ERROR or FATAL. For this, use the macro RCLCPP_[LEVEL] or RCLCPP_[LEVEL]_STREAM to use text streams.
By default, in addition to being sent to /rosout, severity levels INFO or higher will be displayed on the standard output. You can configure the logger to establish another minimum level of severity to be displayed on the standard output:

```
$ ros2 run br2_basics logger --ros-args --log-level debug
```

When there are many nodes in an application, it is recommended to use tools such as rqt_console that allows selecting nodes and severities.
Further readings:

- https://docs.ros.org/en/foxy/Tutorials/Logging-and-logger-configuration.html
- https://docs.ros.org/en/foxy/Concepts/About-Logging.html

The second strategy to iteratively execute a task can be seen in the src/logger_class.cpp program. In addition, we show something widespread in ROS2, which is to implement the nodes inheriting from rclcpp::Node. This approach allows to have a cleaner code and opens the door to many possibilities that will be shown later:

src/logger_class.cpp

```cpp
class LoggerNode : public rclcpp::Node
{
public:
  LoggerNode() : Node("logger_node")
  {
    counter_ = 0;
    timer_ = create_wall_timer(
      500ms, std::bind(&LoggerNode::timer_callback, this));
  }

  void timer_callback()
  {
    RCLCPP_INFO(get_logger(), "Hello %d", counter_++);
  }

private:
  rclcpp::TimerBase::SharedPtr timer_;
  int counter_;
};

int main(int argc, char * argv[]) {
  rclcpp::init(argc, argv);

  auto node = std::make_shared<LoggerNode>();

  rclcpp::spin(node);

  rclcpp::shutdown();
  return 0;
}
```

A timer controls the control loop. This timer produces an event at the desired frequency. When this event happens, it calls the callback that handles it. The advantage is that the node internally adjusts the frequency at which it should be executed without delegating this decision to external code. *Schedule the nodes to know how often they should run.*

To compile these program, the relevant lines in `CMakeLists.txt` are:

- For each executable, an `add_executable` and its corresponding `ament_target_dependencies`.
- An `install` instruction, with all the executables.

CMakeLists.txt

```
add_executable(logger_class src/logger.cpp)
ament_target_dependencies(logger ${dependencies})

add_executable(logger_class src/logger_class.cpp)
ament_target_dependencies(logger_class ${dependencies})

install(TARGETS
  logger
  logger_class
  ...
  ARCHIVE DESTINATION lib
  LIBRARY DESTINATION lib
  RUNTIME DESTINATION lib/${PROJECT_NAME}
)
```

```
$ ros2 run br2_basics logger_class
```

Build the package and run this program to see that the effect is the same as the previous program. Try to modify the frequencies by setting a different time when creating the timer, in `create_wall_timer`.

2.3.2 Publishing and Subscribing

Now extend the node so that, instead of writing a message on the screen, it publishes a message on a topic (Figure 2.8), posting consecutive numbers in a topic called `/counter`. An exploration using the `ros2 interface` command with the `list` and `show` options, lets to find the message that best suits this duty: `std_msgs/msg/Int32`.

It is necessary to include the headers where it is defined to use a message. Since the type of the message to use is `std_msgs/msg/Int32`, notice how from the name of the message we can easily extract which header to include. Just type it, inserting one space before any existing uppercase, and converting all to lowercase. The name of the type is also straightforwards:

```
// For std_msgs/msg/Int32
#include "std_msgs/msg/int32.hpp"

std_msgs::msg::Int32 msg_int32;


// For sensor_msgs/msg/LaserScan
#include "sensor_msgs/msg/laser_scan.hpp"

sensor_msgs::msg::LaserScan msg_laserscan;
```

Focus on the source code of `PublisherNode`:

src/publisher_class.cpp

```
class PublisherNode : public rclcpp::Node
{
public:
  PublisherNode() : Node("publisher_node")
  {
    publisher_ = create_publisher<std_msgs::msg::Int32>("int_topic", 10);
    timer_ = create_wall_timer(
      500ms, std::bind(&PublisherNode::timer_callback, this));
  }

  void timer_callback()
  {
    message_.data += 1;
    publisher_->publish(message_);
  }

private:
  rclcpp::Publisher<std_msgs::msg::Int32>::SharedPtr publisher_;
  rclcpp::TimerBase::SharedPtr timer_;
  std_msgs::msg::Int32 message_;
};
```

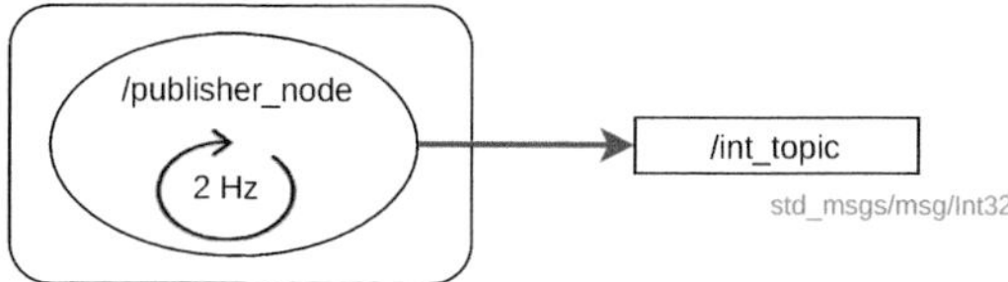

Figure 2.8: Computation Graph for the `Publisher` node.

Let's discuss the important aspects:

- We will use the `std_msgs/msg/Int32` message. From this name, we can deduce that:
  - Its header is `std_msgs/msg/int32.hpp`.
  - The data type is `std_msgs::msg::Int32`.
- Create a publisher, the object in charge of creating the topic (if it does not exist) and publishing the messages. It is possible to obtain more information through this object, such as how many subscribers are on a topic. We use `create_publisher`, which is a public method of `rclcpp::Node`, and it returns a `shared_ptr` to an `rclcpp::Publisher` object. The arguments are the name of the topic and an `rclcpp::QoS` object. This class has a constructor that receives an integer that is the size of the output message queue for that topic so that we can put this size directly, and the C++ compiler will do its magic. We will see later that here we can select different QoS.
- We create a `std_msgs::msg::Int32` message, which we can verify that it only has one data field. Every 500 ms, in the timer callback, we increment the message field and call the publisher's `publish` method to publish the message.

Deeping into: QoS in ROS2

The QoS in ROS2 is an essential and valuable feature in ROS2 and a point of failure, so it must be well understood. In the references at the bottom of this table, you can see what QoS policies can be established and their meaning. The following is an example of how to set QoS policies in C++:

```
publisher = node->create_publisher<std_msgs::msg::String>(
  "chatter", rclcpp::QoS(100).transient_local().best_effort());
```

Default	Reliable	Volatile	Keep Last
Services	Reliable	Volatile	Normal Queue
Sensor	Best Effort	Volatile	Small Queue
DParameters	Reliable	Volatile	Large Queue

Each publisher specifies its QoS, and each publisher can specify its QoS as well. The problem comes because there are QoS that are not compatible, and this will make the subscriber not receive messages:

Compatibility of QoS **durability** profiles		**Subscriber**	
		Volatile	**Transient Local**
Publisher	**Volatile**	Volatile	No Connection
	Transient Local	Volatile	Transient Local

Compatibility of QoS **reliability** profiles		**Subscriber**	
		Best Effort	**Reliable**
Publisher	**Best Effort**	Best Effort	No Connection
	Reliable	Best Effort	Reliable

The criteria really should be that the publisher should have a less restrictive QoS policy than the subscriber. For example, the driver of a sensor should publish its readings with a reliable QoS policy. The subscribers decide if they want the communication to be effectively reliable or prefer Best Effort. In this case, these publishers could be:

```
publisher_ = create_publisher<sensor_msgs::msg::LaserScan>(
  "scan", rclcpp::SensorDataQoS().reliable());
```

and the subscribers could use the same QoS, or remove the reliable part.
Further readings:

- https://docs.ros.org/en/foxy/Concepts/About-Quality-of-Service-Settings.html
- https://design.ros2.org/articles/qos.html
- https://discourse.ros.org/t/about-qos-of-images/18744/16

Run the program:

```
$ ros2 run br2_basics publisher_class
```

And see what we are publishing in the topic:

```
$ ros2 topic echo /int_topic
data: 16
---
data: 17
---
data: 18
...
```

We should see messages with `std_msgs/msg/Int32` messages whose data field is increasing.

Now implement the Node that subscribes to this message:

src/subscriber_class.cpp

```cpp
class SubscriberNode : public rclcpp::Node
{
public:
  SubscriberNode() : Node("subscriber_node")
  {
    subscriber_ = create_subscription<std_msgs::msg::Int32>("int_topic", 10,
      std::bind(&SubscriberNode::callback, this, _1));
  }

  void callback(const std_msgs::msg::Int32::SharedPtr msg)
  {
    RCLCPP_INFO(get_logger(), "Hello %d", msg->data);
  }

private:
  rclcpp::Subscription<std_msgs::msg::Int32>::SharedPtr subscriber_;
};
```

In this code, we have created an `rclcpp::Subscription` to the same topic, with the same type of messages. When creating it, we have indicated that for each message published on this topic, the callback function is called, which receives the message in its `msg` parameter as a `shared_ptr`.

Add this program to `CMakeLists.txt`, build, and run `publisher_class` in one terminal and this program in another, composing the Computation Graph shown in Figure 2.9. We will see how the messages you receive on the topic are displayed on the screen.

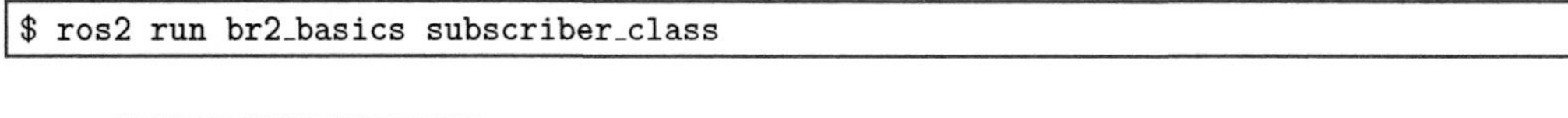

```
$ ros2 run br2_basics subscriber_class
```

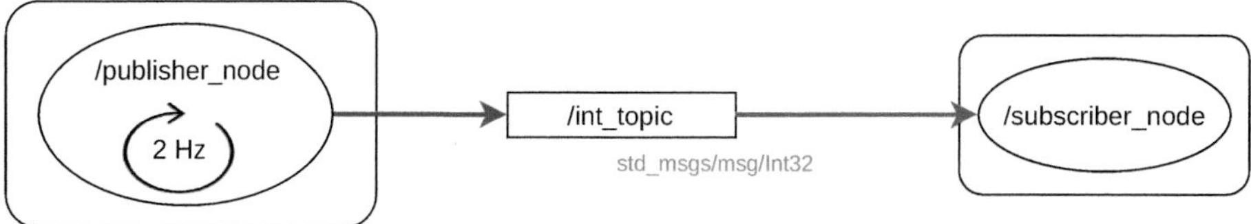

Figure 2.9: Computation Graph for the `Publisher` and `Subscriber` nodes.

2.3.3 Launchers

Up to this point, we have seen that to run a program, we used `ros2 run`. In ROS2, there is also another way to run programs, which is through the command `ros2` launch, and using a file, called launcher, that specifies which programs should be run.

The launcher files are written in Python[1], and their function is declaring which programs to execute with which options or arguments. A launcher can, in turn, include another launcher, allowing you to reuse existing ones.

The need for launchers comes from the fact that a robotic application has many nodes, and they should all be launched simultaneously. Launching one by one and adjusting specific parameters to each one so that the nodes cooperate can be tedious.

Launchers for a package are located in the launch directory of a package, and their name usually ends in _launch.py. Just as ros2 run completed with the programs available in a package, ros2 launch does the same with the available launchers.

From an implementation point of view, a launcher is a python program that contains a generate_launch_description() function that returns a LaunchDescription object. A LaunchDescription object contains actions, among which we highlight:

- **Node** action: to run a program.
- **IncludeLaunchDescription** action: to include another launcher.
- **DeclareLaunchArgument** action: to declare launcher parameters.
- **SetEnvironmentVariable** action: to set an environment variable.

See how we can launch the publisher and subscriber at the same time. Analyze the first launcher in the basics package:

launch/pub_sub_v1_launch.py

```
from launch import LaunchDescription
from launch_ros.actions import Node

def generate_launch_description():
  pub_cmd = Node(
    package='basics',
    executable='publisher',
    output='screen'
  )

  sub_cmd = Node(
    package='basics',
    executable='subscriber_class',
    output='screen'
  )

  ld = LaunchDescription()
  ld.add_action(pub_cmd)
  ld.add_action(sub_cmd)

  return ld
```

There is another implementation alternative of this file in launch/pub_sub_v2_launch.py which behavior is the same. Check it to see the differences. To use launchers, we must install the launchers directory:

[1]Last ROS2 distros lets to create launchers written in Yaml and XML

CMakeLists.txt

```
install(DIRECTORY launch DESTINATION share/${PROJECT_NAME})
```

Build and launch this file:

```
$ ros2 launch br2_basics pub_sub_v2_launch.py
```

In this section, we have seen very simple launchers with very few options. As we progress, we will see more options in increasingly complicated launchers.

2.3.4 Parameters

A node uses the parameters to configure its operation. When your program need configuration files, use parameters. These parameters can be boolean, integer, string, or arrays of any of these types. Parameters are read at run time, usually when a node starts, and their operation depends on these values.

Imagine that a node is in charge of locating a robot using a Particle Filter [9] and requires several parameters, such as a maximum number of particles or the topics from which to receive sensory information. This should not be written in the source code since, if we change the robot or environment, these values may be required to be different.

Look at a node that reads these parameters on startup. Create a `param_reader.cpp` file in the `basics` package:

src/param_reader.cpp

```
class LocalizationNode : public rclcpp::Node
{
public:
  LocalizationNode() : Node("localization_node")
  {
    declare_parameter<int>("number_particles", 200);
    declare_parameter<std::vector<std::string>>("topics", {});
    declare_parameter<std::vector<std::string>>("topic_types", {});

    get_parameter("number_particles", num_particles_);
    RCLCPP_INFO_STREAM(get_logger(), "Number of particles: " << num_particles_);

    get_parameter("topics", topics_);
    get_parameter("topic_types", topic_types_);

    if (topics_.size() != topic_types_.size()) {
      RCLCPP_ERROR(get_logger(), "Number of topics (%zu) != number of types (%zu)",
        topics_.size(), topic_types_.size());
    } else {
      RCLCPP_INFO_STREAM(get_logger(), "Number of topics: " << topics_.size());
      for (size_t i = 0; i < topics_.size(); i++) {
        RCLCPP_INFO_STREAM(
          get_logger(),
          "\t" << topics_[i] << "\t - " << topic_types_[i]);
      }
    }
  }

private:
  int num_particles_;
  std::vector<std::string> topics_;
  std::vector<std::string> topic_types_;
};
```

- All parameters of a node must be declared using methods like `declare_parameter`. In the declaration, we specify the parameter name and the default value.
- We obtain its value with functions like `get_parameter`, specifying the name of the parameter and where to store its value.
- There are methods to do this in blocks.
- The parameters can be read at any time, even subscribe to modifications in real-time. However, reading them to the startup makes your code more predictable.

If we run our program without assigning a value to the parameters, we will see how the default values take value:

```
$ ros2 run br2_basics param_reader
```

Stop executing the program, and execute our program assigning value to one of the parameters. We can do this in setting arguments, starting with `--ros-args`, and -p for setting a parameter:

```
$ ros2 run br2_basics param_reader --ros-args -p number_particles:=300
```

Now pass in values for the remaining parameters. In this case, the two string arrays:

```
$ ros2 run br2_basics param_reader --ros-args -p number_particles:=300
-p topics:= '[scan, image]' -p topic_types:='[sensor_msgs/msg/LaserScan,
sensor_msgs/msg/Image]'
```

If we want to set the parameter values in a launch, we can do it as follows:

launch/param_node_v1_launch.py

```python
from launch import LaunchDescription
from launch_ros.actions import Node

def generate_launch_description():
    param_reader_cmd = Node(
        package='basics',
        executable='param_reader',
        parameters=[{
            'particles': 300,
            'topics': ['scan', 'image'],
            'topic_types': ['sensor_msgs/msg/LaserScan', 'sensor_msgs/msg/Image']
        }],
        output='screen'
    )

    ld = LaunchDescription()
    ld.add_action(param_reader_cmd)

    return ld
```

Although this method may be suitable for assigning values to a few parameters, it is usually convenient to use a file containing the parameters' values with which we want to execute a node. This is the way to have configuration files in ROS2. The chosen format is YAML. Usually, these configuration files are stored in the config directory of our packages, and it is mandatory to mark them to install in the `CMakeLists.txt`, as it was done with the launch directory:

CMakeLists.txt

```
install(DIRECTORY launch config DESTINATION share/${PROJECT_NAME})
```

Let's discuss an important point: what would prevent someone from using a different organization in their packages? Why is the config directory and not set up or startup instead of launch? And why put the source file in another structure? Why use YAML/parameters and not text files or XML and a custom configuration reader? Why use launchers and not a bash script? And why not an application that launches all the necessary nodes?

Of course, a ROS2 developer could make other decisions, but there is a general agreement on doing things. This agreement has the advantage that when another developer tries to use your code, it is much easier to find and identify the critical elements. My recommendation is to follow these conventions. This way, your code can be used by more people, it will be more maintainable in the long term, and receive more collaborations. A company will make this more critical because it will be easier to inherit software when incorporating or replacing developers.

Continue with our example. A file with the parameters with our node could look like this:

config/params.yaml

```
localization_node:
  ros__parameters:
    number_particles: 300
    topics: [scan, image]
    topic_types: [sensor_msgs/msg/LaserScan, sensor_msgs/msg/Image]
```

And execute indicating specifying the location of our file. If we have installed the config directory and compiled it, we can execute:

```
$ ros2 run br2_basics param_reader --ros-args --params-file
install/basics/share/basics/config/params.yaml
```

If we want it to be read in a launcher, we will use:

launch/param_node_v1_launch.py

```
def generate_launch_description():
    ...
    param_reader_cmd = Node(
        package='basics',
        executable='param_reader',
        parameters=[param_file],
        output='screen'
    )
```

2.3.5 Executors

As the nodes in ROS2 are C ++ objects, a process can have more than one node. In fact, in many cases, it can be very beneficial to do so since communications can be accelerated by using shared memory strategies when communication is within the same process. Another benefit is that it can simplify the deployment of nodes if they are all in the same program. The drawback is that a failure in one node could cause all nodes of the same process to terminate.

ROS2 offers you several ways to run multiple nodes in the same process. The most recommended is to make use of the Executors. An Executor is an object to which nodes are added to execute them together. See an example:

Single thread executor

```
int main(int argc, char * argv[]) {
  rclcpp::init(argc, argv);

  auto node_pub = std::make_shared<PublisherNode>();
  auto node_sub = std::make_shared<SubscriberNode>();

  rclcpp::executors::SingleThreadedExecutor executor;

  executor.add_node(node_pub);
  executor.add_node(node_sub);

  executor.spin();

  rclcpp::shutdown();
  return 0;
}
```

Multi thread executor

```
auto node_pub = std::make_shared<PublisherNode>();
auto node_sub = std::make_shared<SubscriberNode>();

rclcpp::executors::MultiThreadedExecutor executor(
      rclcpp::executor::ExecutorArgs(), 8);

executor.add_node(node_pub);
executor.add_node(node_sub);

executor.spin();
}
```

In both codes, we create an executor to which we add the two nodes (Figure 2.10) so that the spin call handles both nodes. The difference between the two is using a single thread for this management, or using eight threads to optimize the processor capabilities.

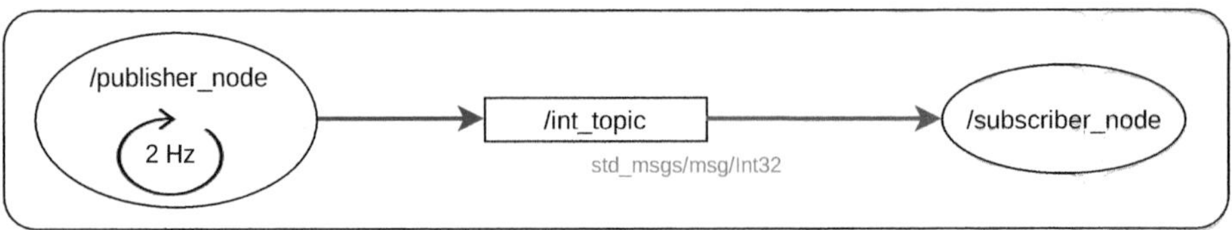

Figure 2.10: Computation Graph for the `Publisher` and `Subscriber` nodes, running in the same process.

2.4 SIMULATED ROBOT SETUP

So far we have seen the basics package, which shows us basic elements of ROS2, how to create nodes, publications, and subscriptions. ROS2 is not a communications middleware, but a robot programming middleware, and this book tries to create behaviors for robots. Therefore, we need a robot. Robots are relatively expensive. It is possible to have a real robot, such as the Kobuki (turtlebot 2) equipped with a laser and an RGBD camera for around 1000€. A robot considered professional can go to several tens of thousands of euros. As not all readers have plans to acquire a robot to run ROS2, we are going to use the Tiago robot in a simulator.

The Tiago robot ("iron" model) from Pal Robotics is formed by a differential base with distance sensors and a torso with an arm, an RGBD camera located on its head.

Among the packages that we have already added to the worskspace, there were already those necessary to simulate the Tiago robot in Gazebo (one of the reference simulator in ROS2). So we will only have to use a launcher that we have created in the `br2_tiago` package:

```
$ ros2 launch br2_tiago sim.launch.py
```

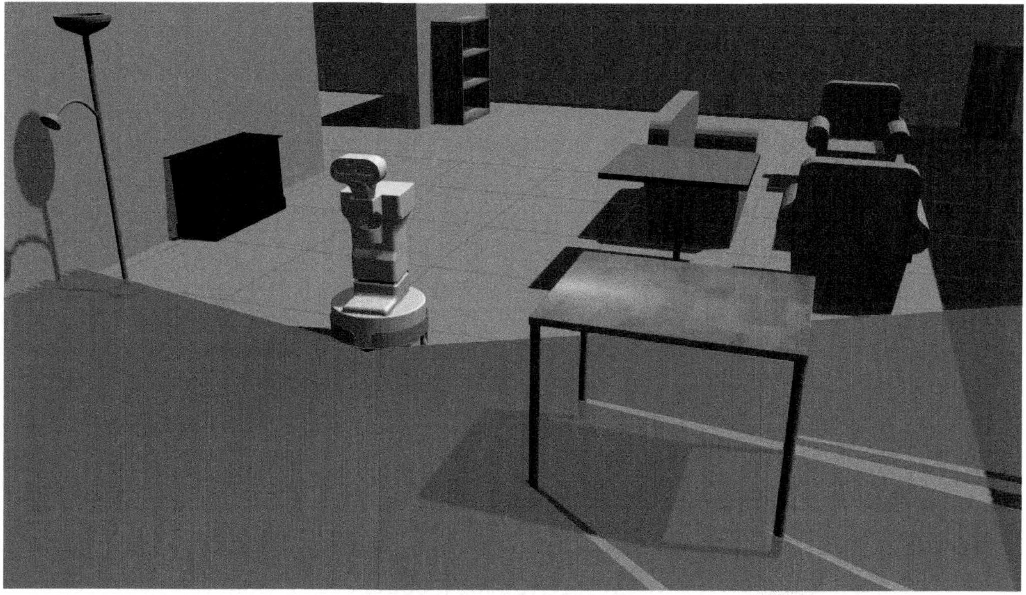

Figure 2.11: Simulating Tiago robot in Gazebo.

There are several worlds available (you can examine `ThirdParty/pal_gazebo_worlds/worlds`). By default, the world that is loaded is `home.world`. If you want to use a different one, you can use the launcher `world` parameter, as shown in the following examples:

```
$ ros2 launch br2_tiago sim.launch.py world:=factory
$ ros2 launch br2_tiago sim.launch.py world:=featured
$ ros2 launch br2_tiago sim.launch.py world:=pal_office
$ ros2 launch br2_tiago sim.launch.py world:=small_factory
$ ros2 launch br2_tiago sim.launch.py world:=small_office
$ ros2 launch br2_tiago sim.launch.py world:=willow_garage
```

One of the first things you can do when you use a robot for the first time and have just launched its driver or simulation is to see what topics it provides, either as a publisher or a subscriber. That will be the interface we will use to receive information from the robot and send it commands. Open a new terminal and execute:

```
$ ros2 topic list
```

This will be the main interface with the robot's sensors and actuators. Figure 2.14 shows a non-exhaustive way the nodes and topics that are available to the programmer to interact with the simulated robot:

- Virtually all nodes are within the Gazebo simulator process. Outside there are only two of them:

  /twist_mux Create several subscribers to topics that receive robot speeds, but from different sources (mobile, tablet, keys, navigation, among others).

  /robot_state_publisher It is a standard node in ROS2 that reads the description of a robot from a URDF file and subscribes to the status of each of the robot's joints. In addition to publishing this description in URDF, it creates and updates the robot frames in the TFs system (we will explain the TF system in the next chapters), a system to represent and link the different geometric axes of reference in the robot.

- The nodes on the left take care of the sensors. They publish information from the robot's camera, imu, laser, and sonar. The most complex node is the camera node, an RGBD sensor, since it publishes the depth and RGB images separately. Each image has associated a topic `camera_info` that contains the intrinsic values of the robot's camera. For each sensor, the standard message types are used for the information provided.

- The nodes at the bottom use the same interface to move the head and the torso. They all use the `joint_trajectory_controller` from the `ros2_control` package.

- The nodes on the right are responsible for the following:

  /joint_state_broadcaster Publish the status of each of the joints of the robot.

  /mobile_base_controller Makes the robot base move with the speed commands it receives. In addition, it publishes the estimated displacement of the base.

First, teleoperate the robot to move it. For this, ROS2 has several packages that take commands from the keyboard, from a PS or XBox controller, or a mobile phone, and publish `geometry_msgs/msg/Twist` messages in a topic. In this case we will use `teleop_twist_keyboard`. This program receives keystrokes by stdin in publishes `/cmd_vel` movement commands.

As the topic `/cmd_vel` of `teleop_twist_keyboard` does not match any input topic of our robot, we must do a remap. A remap (Figure 2.12) allows you to change the name of one of its topics when executing (at *deployment time*). In this case, we are going to execute `teleop_twist_keyboard` indicating that instead of publishing in `/cmd_vel`, publish in the topic `/key_vel` of the robot:

```
$ ros2 run teleop_twist_keyboard teleop_twist_keyboard --ros-args -r
cmd_vel:=key_vel
```

Now we can use the keys indicated by `teleop_twist_keyboard` to move the robot.

Remapping a topic is an important feature of ROS2 that allows different ROS2 programs from other developers to work together.

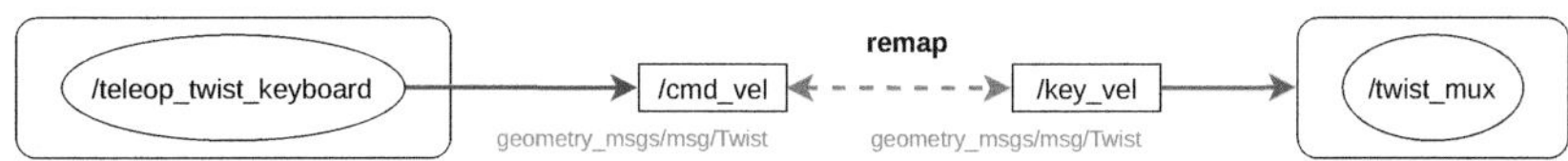

Figure 2.12: Connection between the Tiago and the teleoperator velocity topics, using a remap.

Now is the time to see the robot's sensory information. Until now, we could use `ros2 topic` to see the topics of the camera or the laser with one of these commands:

```
$ ros2 topic echo /scan_raw

$ ros2 topic echo /head_front_camera/rgb/image_raw
```

But it is hard to show sensory information, especially if it is so complex. Use the `--no-arr` option so that it does not display the content of the data arrays.

```
$ ros2 topic echo --no-arr /scan_raw

$ ros2 topic echo --no-arr /head_front_camera/rgb/image_raw
```

Analyze the information it shows. There is a common field in both messages, which is common in messages with perceptual information and is repeated in many types of messages, especially those that end in the adjective "*Stamped". It has a header of type `std_msgs/msg/Header`. As we have just seen, Messages can be defined by composing basic types (int32, Float64, String) or already existing messages, like this one.

The header is tremendously helpful for handling sensory information in ROS2. When a sensor driver publishes messages with its data, it uses the header to tag this reading with:

- The data capture timestamp. Even if a message is received or processed late, the reading can be placed at its corresponding capture moment, supporting some latencies.

- The frame in which it was taken. A frame is an axis of references in which the spatial information (coordinates, distances, etc) contained in the message makes sense. Usually, each sensor has its frame (even several).

A robot is geometrically modeled using a tree whose tree nodes are the frames of a robot. By convention, a frame should have a single parent frame and all required child frames. The parent-child relationship is through a geometric transformation that includes a translation and a rotation. The frames usually appear at points on the robot subject to variation, as in the case of the motors joining the robot's parts.

ROS2 has a system called TF, which we will explain in next chapters, which maintains these relationships through two topics `/tf`, for geometric transformations that vary, and `/tf_static` if they are fixed.

ROS2 has several tools that help us display sensory and geometric information, and perhaps the most popular is RViz2. Start by running it by typing in a terminal:

```
$ ros2 run rviz2 rviz2
```

RViz2 is a viewer that allows to display information contained in the topics. If this is your first time opening RViz2, it will probably appear quite empty; only a grid through which we can navigate using the keys and the mouse. We will discover the information about our robot step by step, as shown in Figure 2.13:

1. On the left, in the Displays panel, RViz2 has some global options in which we have to specify what our Fixed Frame is, that is, the coordinate axis of the 3D visualization shown on the right. For now, we are going to select `base_footprint`. By convention in ROS2, this frame is a frame that is in the center of the robot, on the ground, and is a good starting point for our exploration.

2. In the Displays Panel, we are going to add different visualizations. The first will be to see the frames of the robot. Press the Add button, and look in the "By display type" tab, the TF element. All the robot frames will appear instantly. If they seem like a lot to you, display the TF component in the Displays Panel, uncheck the "All Enabled" box and start adding or removing the frames you want.

3. Add several elements to Gazebo, as seen in Figure below. If not, we will not perceive much either.

4. Add the laser information of the robot. Press Add again and in the "By Topic" tab, select the topic `/scan_raw`, which already indicates that it is a LaserScan. In the LaserScan element that has been added in the Display Panel, we can see information and change the display options. Display the options for this element:

    - The Status should show ok and have a counter that goes up as it receives messages. If it displays an error, it usually contains information that can help us figure out how to fix it.

- The Topic has to do with the topic to show and the QoS with which RViz2 subscribes to that topic. If we do not see anything, it may be that we have not selected a QoS compatible.
- From here on, the rest of the options are specific to this type of message. We can change the size of the dots that represent laser readings, their color, or even the visual element used.

5. As done with the laser, add a visualization of the topic that contains the PointCloud2 (`head_front_camera` → `depth_registered` → `points`).

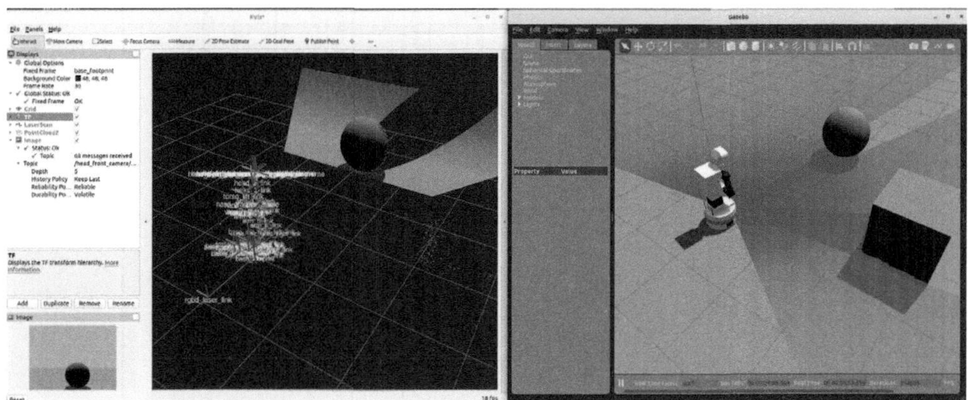

Figure 2.13: RViz2 visualizing TFs and the sensor information of the Tiago robot.

Use the teleoperator to move the robot. In RViz2, the movement of the robot is not appreciated, only the frame `odom` moving. This is because the center of this visualization is always the Fixed Frame, which we now have as `base_footprint`. Change the Fixed Frame to `odom`, which is a Frame that represents the position of the robot when it started. Now we can appreciate the robot's movement around its environment. The `odom` → `base_footprint` transform, by convention in ROS2, collects the translation and rotation calculated by the robot driver from its starting point.

With this, we have explored the capabilities of the simulator robot and various tools for managing our robot.

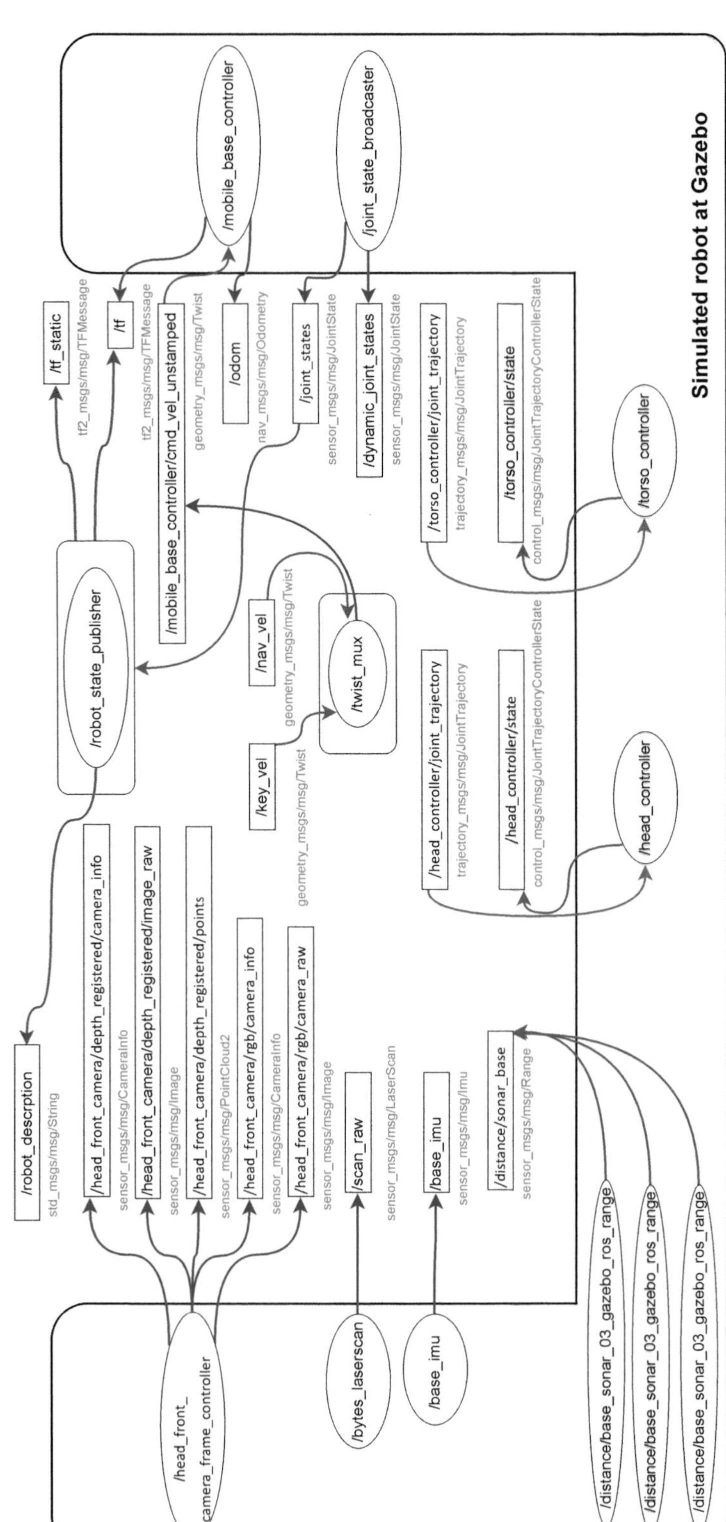

Figure 2.14: Computation Graph for the Tiago robot, displaying the relevant topics.

## CHAPTER 3

First Behavior: Avoiding Obstacles with Finite States Machines

THIS section aims to apply everything shown until now to create seemingly "smart" behavior. This exercise will put together many things we have presented and show how effective it is to program a robot using ROS2. In addition, we will address some issues in robot programming.

The *Bump and Go* behavior uses the robot's sensor to detect nearby obstacles in front of the robot. The robot moves forward, and when it detects an obstacle, it goes back and turns for a fixed time to move forward again. Although it is a simple behavior, some decision-making approach is recommended since our code, even if it is simple, can start to grow out of order as we solve the problems that may arise. In this case, we will use a Finite State Machine (FSM).

An FSM is a mathematical computational model that we can use to define the behavior of a robot. It is made up of states and transitions. A robot keeps producing an output in one state until the condition of an outgoing transition is fulfilled and it transits to the target state of this transition.

Applying an FSM can significantly reduce the complexity of solving a problem when we implement simple behaviors. For a moment, try to think about how to approach the *Bump and Go* problem using loops, ifs, temporary variables, counters, timers. It would be a complex program to understand and follow its logic. Once finished, adding some additional conditions will probably make to throw away what we have done and start over.

Applying an FSM-based solution to the *Bump and Go* problem is straightforward. Think about the different outputs that the robot must produce (stop, move forward, go back, and turn). Each of these actions will have its own state. Now think about the transitions between states (connection and condition), and we will obtain an FSM like the one shown in Figure 3.1.

DOI: 10.1201/9781003289623-3

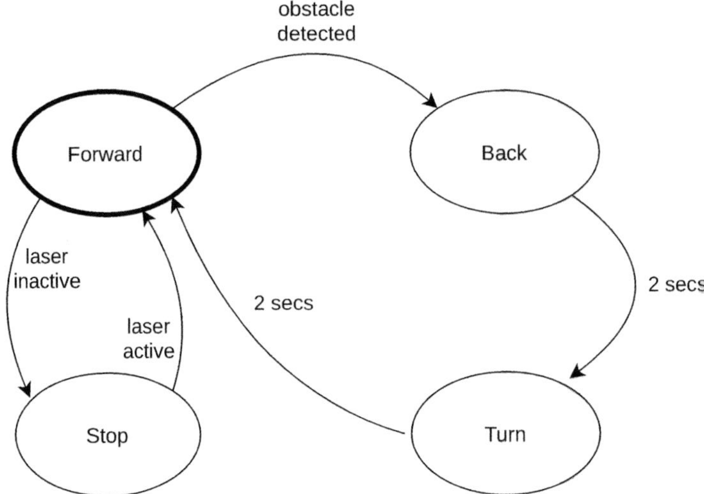

Figure 3.1: States and Transitions for solving the *Bump and Go* problem using a FSM.

3.1 PERCEPTION AND ACTUATION MODELS

This section analyzes what perceptions we use to solve the problem and what actions we can produce.

In both models, first of all, we must define the used geometric conventions:

- ROS2 uses the metric International System of Measurements (SI). For different dimensions, we will consider the units of meters, seconds, and radians. Linear speeds should be m/s, rotational speeds rad/s, linear accelerations m/s^2, and so on.

- In ROS2 we are right-handed (left part of Figure 3.2): x grows forward, y to the left, and z grows up. If we establish the reference origin on our chest, a coordinate whose x is negative would be behind us, and a positive z would be above us.

- Angles are defined as rotations around the axes. Rotation around x is sometimes called the roll, y pitch, and z yaw.

- Angles grow by turning to the left (right part of Figure 3.2). Angle 0 is forward, π is back, and $\pi/2$ is left.

In this problem, we will use the information of the laser sensor, which we saw in the previous chapter that was in the topic `/scan_raw`, and whose type was `sensor_msgs/msg/LaserScan`. Check this message format by typing:

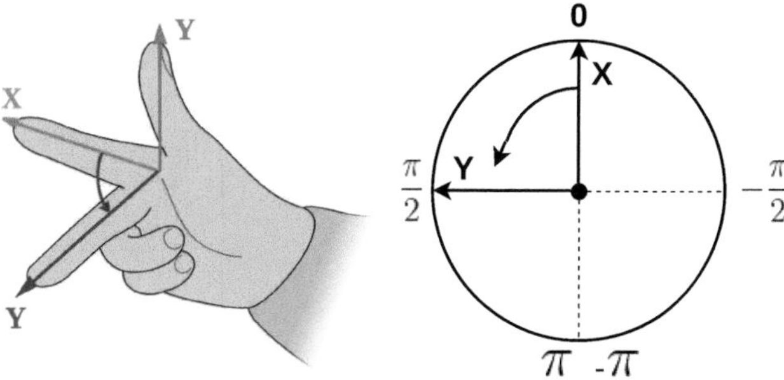

Figure 3.2: Axis and angles conventions in ROS.

```
$ ros2 interface show /sensor_msgs/msg/LaserScan

# Single scan from a planar laser range-finder
#
# If you have another ranging device with different behavior (e.g. a sonar
# array), please find or create a different message, since applications
# will make fairly laser-specific assumptions about this data

std_msgs/Header header  # timestamp in the header is the acquisition time of
                        # the first ray in the scan.
                        #
                        # in frame frame_id, angles are measured around
                        # the positive Z axis (counterclockwise, if Z is up)
                        # with zero angle being forward along the x axis

float32 angle_min       # start angle of the scan [rad]
float32 angle_max       # end angle of the scan [rad]
float32 angle_increment # angular distance between measurements [rad]

float32 time_increment  # time between measurements [seconds] - if your scanner
                        # is moving, this will be used in interpolating pos
                        # of 3d points
float32 scan_time       # time between scans [seconds]

float32 range_min       # minimum range value [m]
float32 range_max       # maximum range value [m]

float32[] ranges        # range data [m]
                        # (Note: values < range_min or > range_max should be
                        # discarded)
float32[] intensities   # intensity data [device-specific units]. If your
                        # device does not provide intensities, please leave
                        # the array empty.
```

To see one of these laser messages (without showing the content of the readings), launch the simulator and type:

```
$ ros2 topic echo /scan_raw --no-arr
---
header:
  stamp:
    sec: 11071
    nanosec: 445000000
  frame_id: base_laser_link
angle_min: -1.9198600053787231
angle_max: 1.9198600053787231
angle_increment: 0.005774015095084906
time_increment: 0.0
scan_time: 0.0
range_min: 0.05000000074505806
range_max: 25.0
ranges: '<sequence type: float, length: 666>'
intensities: '<sequence type: float, length: 666>'
---
```

In figure 3.3 we can see the interpretation of this message. The key is that in the **ranges** field are the distances to obstacles. Position 0 of this `std::vector` (arrays in messages are represented as `std::vector` in C++) corresponds to angle −1.9198, position 1 is this angle plus the increment, until this vector is completed. It is easy to check that if we divide the range (maximum angle minus minimum angle) by the increment, we get these 666 readings, which is the size of the ranges vector.

Most messages, especially if they contain spatially interpretable information, have a header containing the timestamp and the sensor frame. Note that a sensor can be mounted in any position on the robot and any orientation, even in some moving parts. The sensor frame must have a geometric connection (a rotation and translation) to the rest. On many occasions, we will need to transform the coordinates of the sensory information to the same frame to fuse it, which is usually `base_footprint` (the center of the robot, at ground level, pointing forward). These geometric manipulations are explained in next chapter.

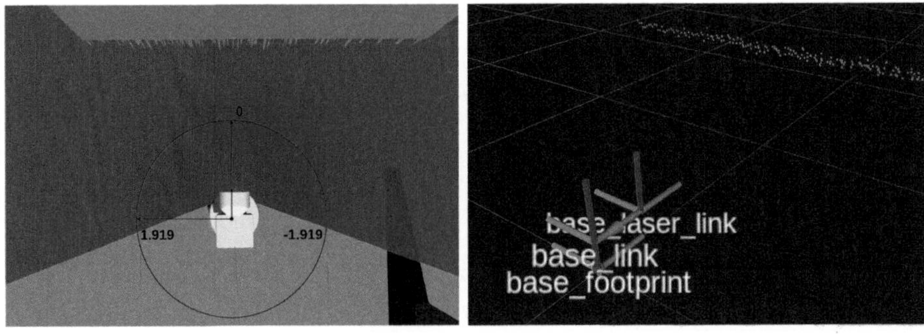

Figure 3.3: Laser scan interpretation in the simulated Tiago (left). Laser frame with respect to other main frames (right).

In our problem, we are only interested in whether there is an obstacle in front of the robot, which is angle 0, and this corresponds exactly to the content of the middle position of the vector of ranges. We can use the original frame of the sensor since it is aligned, a little forward and up, with `base_footprint`.

An essential feature of ROS is standardization. Once a consensus has been reached in the community on the format in which the information produced by a laser sensor is encoded, all laser driver developers should use this format. This consensus means that the message format must be general enough to support any laser sensor. In the same way, an application developer must exploit the information in this message for his program to function correctly regardless of the characteristics of the sensor that produced the sensory reading. The great advantage of this approach is that we can make any ROS program work with any ROS-supported laser, allowing the software to be truly portable between robots. Also, an experienced ROS developer does not have to learn new, manufacturer-defined formats. Finally, using this format puts at your disposal a wide variety of utilities to filter or monitor laser information. This approach applies to all types of sensors and actuators in ROS, which may be one reason for the success of this framework.

Regarding the **action model** in this problem, we will send the robot translation and rotation speeds to topic `/nav_vel`, which is of type `geometry_msgs/msg/Twist`. Let's see this message format:

```
$ ros2 interface show geometry_msgs/msg/Twist

Vector3 linear
Vector3 angular

$ ros2 interface show geometry_msgs/msg/Vector3

float64 x
float64 y
float64 z
```

All robots use this message format to receive speeds, allowing generic teleoperation programs (with keyboard, joystick, mobiles, etc.) and navigation in ROS. Once again, we are talking about standardization.

The `geometry_msgs/msg/Twist` message is much more generic than what our robot supports. We cannot make it move in Z (it cannot fly) or move laterally with just two wheels. It is a differential robot. We could probably do more translations and rotations if we had a quadcopter. We can only make it go forward or backward, rotate, or combine both. For this reason, we can only use the fields `linear.x` and `angular.z` (rotation to the Z-axis, positive velocities to the left, as indicated in Figure 3.2).

3.2 COMPUTATION GRAPH

The Computation Graph of this application will be pretty simple: A single node that subscribes to the laser topic publishes speed commands to the robot.

The control logic interprets the input sensory information and produces the control commands. This logic is what we are going to implement with an FSM. The logic control will run iteratively at 20 Hz. The execution frequencies depend on publishing the control commands. If it is not published above 20 Hz, some robots stop, which is very convenient so that there are no robots without control in the laboratory.

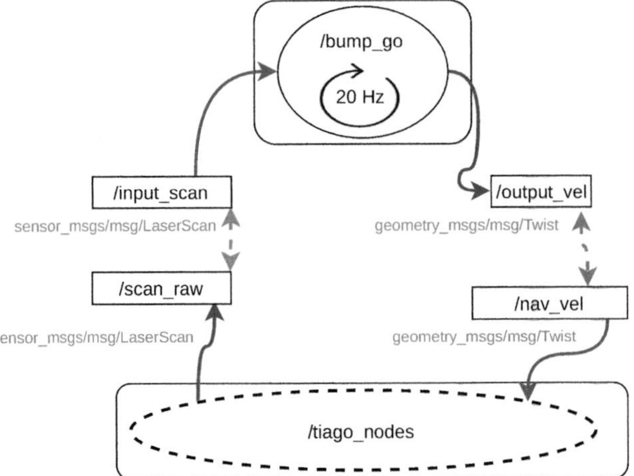

Figure 3.4: Computation Graph for *Bump and Go* project.

Commonly, the frequency at which we receive information is not the same as the frequency we must publish it. You have to deal with this. *Engineers do not complain about problem conditions – they fix them.*

If we want our software to run on different robots, we must not specify specific topics for a robot. In our case, the topic that it subscribes to is `/input_scan`, and it publishes in `/output_vel`. These topics do not exist or correspond to those of our simulated robot. When executing it (at deployment), we will remap the ports to connect them to the real topics of the specific robot.

Let's discuss a point here. Why are we using remaps instead of passing the name of the topics as parameters? Well, it is an alternative that many ROS2 developers advocate. Perhaps this alternative is more convenient when a node does not always have the same subscribers/publishers, and this can only be specified in a YAML file of configuration parameters.

A good approach is that if the number of publishers and subscribers in a node is known, use generic topic names, like the ones used in this example, and perform a remap. It may even be better to use common topic names (`/cmd_vel` is a common topic for many robots). A seasoned ROS2 programmer will read in the documentation what topics it uses, find out with a `ros2 node info`, and quickly make it work with remaps, instead of looking for the correct parameters to be set up in the configuration files.

Although this book primarily uses C++, in this chapter we will provide two similar implementations, one in C++ and other in Python, each in different packages: `br2_fsm_bumpgo_cpp` and `br2_fsm_bumpgo_py`. Both are already in the workspace created in previous chapters and the annex to this book. Let's start with the C++ implementation.

3.3 *BUMP AND GO* IN C++

The br2_fsm_bumpgo_cpp package has the following structure:

Package br2_fsm_bumpgo_cpp

```
br2_fsm_bumpgo_cpp
├── CMakeLists.txt
├── include
│   └── br2_fsm_bumpgo_cpp
│       └── BumpGoNode.hpp
├── launch
│   └── bump_and_go.launch.py
├── package.xml
└── src
    ├── br2_fsm_bumpgo_cpp
    │   └── BumpGoNode.cpp
    └── bumpgo_main.cpp
```

The usual way for nodes to be implemented as classes that inherit from rclcpp::Node, separating declaration and definition, within a namespace that matches the package name. In our case, the definition (BumpGoNode.cpp) will be in src/br2_fsm_bumpgo_cpp, and the header (BumpGoNode.hpp) will be in include/br2_fsm_bumpgo_cpp. In this way, we separate the implementation of the programs from the implementation of the nodes. This strategy allows having several programs with different strategies for creating nodes. The main program, whose function is to instantiate the node and call to the **spin()** function, is in src/bumpgo_main.cpp. We have also included a launcher (launch/bump_and_go.launch.py) to facilitate its execution.

In this book, we will analyze partial pieces of the code of a package, focusing on different concrete aspects to teach interesting concepts. We will not exhaustively show all the code since the reader has it available in his workspace, the repository, and the annexes.

3.3.1 Execution Control

The node execution model consists of calling the control_cycle method at a frequency of 20 Hz. For this, we declare a timer and start it in the constructor to call the control_cycle method every 50 ms. The control logic, implemented with an FSM, will publish the commands in speeds.

include/bump_go_cpp/BumpGoNode.hpp

```cpp
class BumpGoNode : public rclcpp::Node
{
  ...
private:
  void scan_callback(sensor_msgs::msg::LaserScan::UniquePtr msg);
  void control_cycle();

  rclcpp::Publisher<geometry_msgs::msg::Twist>::SharedPtr vel_pub_;
  rclcpp::Subscription<sensor_msgs::msg::LaserScan>::SharedPtr scan_sub_;
  rclcpp::TimerBase::SharedPtr timer_;

  sensor_msgs::msg::LaserScan::UniquePtr last_scan_;
};
```

Look at the detail of the laser callback header. We have used `UniquePtr` (an alias for `std::unique_ptr`) instead of `SharedPtr`, as we have seen so far. The Callbacks in ROS2 can have different signatures, depending on the needs. These are different alternatives for the callbacks:

```
1. void scan_callback(const sensor_msgs::msg::LaserScan & msg);
2. void scan_callback(sensor_msgs::msg::LaserScan::UniquePtr msg);
3. void scan_callback(sensor_msgs::msg::LaserScan::SharedConstPtr msg);
4. void scan_callback(const sensor_msgs::msg::LaserScan::SharedConstPtr & msg);
5. void scan_callback(sensor_msgs::msg::LaserScan::SharedPtr msg);
```

Some other signatures allow to obtain information about the message (timestamp in origin and destination, and identifier of the sender) and even the serialized message, but that is only used in very specialized cases.

Up to this point, we had used signature 1, but now we use signature 2. Check out the implementation of the laser callback at `scan_callback`. Instead of making a copy of the message (which could be computationally expensive for large messages) or sharing the pointer, we will acquire this message in property, and we will store the reference to the data in `last_scan_`. This way, rclcpp queues will no longer need to manage their lifecycle, saving time. We recommend using `UniquePtr` when possible to improve the performance of your nodes.

src/bump_go_cpp/BumpGoNode.cpp

```
BumpGoNode::BumpGoNode()
: Node("bump_go")
{
  scan_sub_ = create_subscription<sensor_msgs::msg::LaserScan>(
    "input_scan", rclcpp::SensorDataQoS(),
    std::bind(&BumpGoNode::scan_callback, this, _1));

  vel_pub_ = create_publisher<geometry_msgs::msg::Twist>("output_vel", 10);
  timer_ = create_wall_timer(50ms, std::bind(&BumpGoNode::control_cycle, this));
}

void
BumpGoNode::scan_callback(sensor_msgs::msg::LaserScan::UniquePtr msg)
{
  last_scan_ = std::move(msg);
}

void
BumpGoNode::control_cycle()
{
  // Do nothing until the first sensor read
  if (last_scan_ == nullptr)
    return;

    vel_pub_->publish(...);
}
```

Another noteworthy detail about this constructor is that the publication use the default QoS, which is reliable + volatile. In case of subscriptions, we will use `rclcpp::SensorDataQoS()` (a packed QoS definition using best effort, volatile, and appropiate queue size for sensors).

As a general rule, for a communication to be compatible, the quality of service of the publisher should be reliable, and it is the subscriber who can choose to relax it to be the best effort. When creating sensor drivers, publishing their readings using `rclcpp::SensorDataQoS()` is not a good idea because if a subscriber requires reliable QoS and publisher is best effort, communication will fail.

Finally, the first thing to do in `control_cycle` is to check if `last_scan_` is valid. This method may be executed before the first message arrives with a laser scan. In this case, this iteration is skipped.

3.3.2 Implementing a FSM

Implementing an FSM in a C++ class is not complicated. It is enough to have a member variable `state_` that stores the current state, which we can encode as a constant or an enum. In addition, it is helpful to have a variable `state_ts_` that indicates the time when transit to the current state, allowing to transit from states using timeouts.

include/bump_go_cpp/BumpGoNode.hpp

```
class BumpGoNode : public rclcpp::Node
{
  ...
private:
  void control_cycle();

  static const int FORWARD = 0;
  static const int BACK = 1;
  static const int TURN = 2;
  static const int STOP = 3;
  int state_;
  rclcpp::Time state_ts_;
};
```

Remember that the control logic is in method `control_cycle`, which runs at 20 Hz. There can be no infinite loops or long waits in this method. It must be designed to iteratively call this method to accomplish its task.

Control logic is typically implemented with a `switch` statement, with a state in each case. In the following code, we have only shown the case of the `FORWARD` state. There is also a structure in this case: first, the output computation in the current state (setting speeds to publish) and then check every transition condition. If any returns true (the condition is met), the `state_` is set to the new state and `state_ts_` is updated.

When declaring a message type variable, all its fields are set by default to their default value, or 0 or empty depending on their type. That is why in the complete code, we only assign the field that is not 0.

src/bump_go_cpp/BumpGoNode.cpp

```
BumpGoNode::BumpGoNode()
: Node("bump_go"),
  state_(FORWARD)
{
  ...
  state_ts_ = now();
}

void
BumpGoNode::control_cycle()
{
  switch (state_) {
    case FORWARD:

      // Do whatever you should do in this state.
      // In this case, set the output speed.

      // Checking the condition to go to another state in the next iteration
      if (check_forward_2_stop())
        go_state(STOP);
      if (check_forward_2_back())
        go_state(BACK);

      break;
      ...
  }
}

void
BumpGoNode::go_state(int new_state)
{
  state_ = new_state;
  state_ts_ = now();
}
```

Look at three methods with interesting code from the implementation point of view. The first is the code of the `forward` → `back` transition that checks if there is an obstacle in front of the robot. As we said before, this is done by accessing the central element of the vector that contains the distances in the laser reading:

src/bump_go_cpp/BumpGoNode.cpp

```
bool
BumpGoNode::check_forward_2_back()
{
  // going forward when deteting an obstacle
  // at 0.5 meters with the front laser read
  size_t pos = last_scan_->ranges.size() / 2;
  return last_scan_->ranges[pos] < OBSTACLE_DISTANCE;
}
```

The second interesting snippet is the transition from `forward` → `stop` when the last laser read is considered too old. The `now` method of `rclcpp::Node` returns the current time as an `rclcpp::Time`. From the time that is in the header of the last reading we can create another `rclcpp::Time`. Its difference is a `rclcpp::Duration`. To make comparisons, we can use its `seconds` method, which returns the time in seconds as a double, or we can, as we have done, directly compare it with another `rclcpp::Duration`.

src/bump_go_cpp/BumpGoNode.cpp

```
bool
BumpGoNode::check_forward_2_stop()
{
  // Stop if no sensor readings for 1 second
  auto elapsed = now() - rclcpp::Time(last_scan_->header.stamp);
  return elapsed > SCAN_TIMEOUT;
}
```

The last snippet is similar to the previous one, but now we take advantage of having the `state_ts_` variable updated, and we can transition from `back` → `turn` after 2 s.

src/bump_go_cpp/BumpGoNode.cpp

```
bool
BumpGoNode::check_back_2_turn()
{
  // Going back for 2 seconds
  return (now() - state_ts_) > BACKING_TIME;
}
```

3.3.3 Running the Code

So far, we have limited ourselves to the class that implements the `BumpGoNode` node. Now we have to see where we create an object of this class to execute it. We do this in the main program that creates a node and passes it to a blocking call to `rclcpp::spin` that will manage the messages and timer events calling to their callbacks.

src/bumpgo_main.cpp

```
int main(int argc, char * argv[])
{
  rclcpp::init(argc, argv);

  auto bumpgo_node = std::make_shared<br2_fsm_bumpgo_cpp::BumpGoNode>();
  rclcpp::spin(bumpgo_node);

  rclcpp::shutdown();

  return 0;
}
```

Now run the program. Open a terminal to run the simulator:

```
$ ros2 launch br2_tiago sim.launch.py
```

Next, Open another terminal and run the program, taking into account that there are arguments to specify in the command line:

- Remap `input_scan` to `/scan_raw`, and `ouput_vel` to `/nav_vel` (`-r` option).
- When using a simulator, set the `use_sim_time` parameter to `true`. This causes the time to be taken from the topic `/clock`, published by the simulator, instead of the current one from the computer.

```
$ ros2 run br2_fsm_bumpgo_cpp bumpgo --ros-args -r output_vel:=/nav_vel -r
input_scan:=/scan_raw -p use_sim_time:=true
```

See how the robot moves forward until it detects an obstacle then does an avoidance maneuver.

Because it is tedious to put so many remapping arguments in the command line, we have created a launcher that specifies the necessary arguments and remaps to the node.

launch/bump_and_go.launch.py

```python
bumpgo_cmd = Node(package='br2_fsm_bumpgo_cpp',
    executable='bumpgo',
    output='screen',
    parameters=[{
      'use_sim_time': True
    }],
    remappings=[
      ('input_scan', '/scan_raw'),
      ('output_vel', '/nav_vel')
    ])
```

Use this launcher instead of the last `ros2 run`, only by typing:

```
$ ros2 launch br2_fsm_bumpgo_cpp bump_and_go.launch.py
```

3.4 *BUMP AND GO* BEHAVIOR IN PYTHON

In addition to C++, Python is one of the languages officially supported in ROS2 through the rclpy client library. This section will reproduce what we have done in the previous section, but with Python. Verify by comparison the differences and similarities in the development of both languages. Also, once the principles of ROS2 have been explained throughout the previous chapters, the reader will recognize the elements of ROS2 in Python code, as the principles are the same.

Although we provided the complete package, if we had wanted to create a package from scratch, we could have used the `ros2 pkg` command to create a skeleton.

```
$ ros2 pkg create --build-type ament_python br2_fsm_bumpgo_py --dependencies
sensor_msgs geometry_msgs
```

As it is a ROS2 package, there is still a `package.xml` similar to the C++ version, but there is no longer a `CMakeLists.txt`, but a `setup.cfg` and `setup.py`, typical of Python packages that use distutils[1].

At the root of this package, there is a homonymous directory that only has a file `__init__.py` which indicates that there will be files with Python code. Let's create the file `bump_go_main.py` there. While in C++, it is common and convenient to separate the source code into several files. In this case, everything is in the same file.

[1] https://docs.python.org/3/library/distutils.html

3.4.1 Execution Control

As in the previous example, we will first show the code ignoring the details of the behavior, only those related to the ROS2 concepts to handle:

bump_go_py/bump_go_main.py

```
import rclpy

from rclpy.duration import Duration
from rclpy.node import Node
from rclpy.qos import qos_profile_sensor_data
from rclpy.time import Time

from geometry_msgs.msg import Twist
from sensor_msgs.msg import LaserScan

class BumpGoNode(Node):
    def __init__(self):
        super().__init__('bump_go')

        ...

        self.last_scan = None
        self.scan_sub = self.create_subscription(
            LaserScan,
            'input_scan',
            self.scan_callback,
            qos_profile_sensor_data)

        self.vel_pub = self.create_publisher(Twist, 'output_vel', 10)
        self.timer = self.create_timer(0.05, self.control_cycle)

    def scan_callback(self, msg):
        self.last_scan = msg

    def control_cycle(self):
        if self.last_scan is None:
            return

        out_vel = Twist()

        # FSM

        self.vel_pub.publish(out_vel)

def main(args=None):
    rclpy.init(args=args)

    bump_go_node = BumpGoNode()

    rclpy.spin(bump_go_node)

    bump_go_node.destroy_node()
    rclpy.shutdown()

if __name__ == '__main__':
    main()
```

Recall that the goal is to create a node that subscribes to the laser readings and issues speed commands. The control cycle executes at 20 Hz to calculate the robot control based on the last reading received. Therefore, our code will have a subscriber, a publisher, and a timer.

This code is similar to the one developed in C++: define a class that inherits from Node, and in the main, it is instantiated and called spin with it. Let's see some details:

- Inheriting from Node, we call the base class constructor to assign the node name. The Node class and all associated data types (Time, Duration, QoS,...) are in rclpy, imported at startup, and these items separately.
- The types of messages are also imported, as seen in the initial part.
- We create the publisher, the subscriber, and the timer in the constructor. Note that the API is practically similar to C++. Also, in Python, we can access predefined qualities of service (`qos_profile_sensor_data`).
- In the callback of the laser messages, we store the last message received in the variable `self.last_scan`, which was initialized to `None` in the constructor. In this way, verify in the control cycle (`control_cycle`) that no laser reading has reached us.

3.4.2 Implementing the FSM

The direct translation of the FSM in C++ from the previous section to Python has nothing interesting. The only detail is that to obtain the current time, we have to ask for the clock first through the `get_clock` method:

bump_go_py/bump_go_main.py

```
class BumpGoNode(Node):
    def __init__(self):
        super().__init__('bump_go')

        self.FORWARD = 0
        self.BACK = 1
        self.TURN = 2
        self.STOP = 3
        self.state = self.FORWARD
        self.state_ts = self.get_clock().now()

    def control_cycle(self):

        if self.state == self.FORWARD:
          out_vel.linear.x = self.SPEED_LINEAR

          if self.check_forward_2_stop():
            self.go_state(self.STOP)
          if self.check_forward_2_back():
            self.go_state(self.BACK)

        self.vel_pub.publish(out_vel)

    def go_state(self, new_state):
        self.state = new_state
        self.state_ts = self.get_clock().now()
```

Perhaps the most remarkable aspect in this code, similar to its version in C++, is the treatment of time and durations:

bump_go_py/bump_go_main.py

```
    def check_forward_2_back(self):
        pos = round(len(self.last_scan.ranges) / 2)
        return self.last_scan.ranges[pos] < self.OBSTACLE_DISTANCE

    def check_forward_2_stop(self):
        elapsed = self.get_clock().now() - Time.from_msg(self.last_scan.header.stamp)
        return elapsed > Duration(seconds=self.SCAN_TIMEOUT)

    def check_back_2_turn(self):
        elapsed = self.get_clock().now() - self.state_ts
        return elapsed > Duration(seconds=self.BACKING_TIME)
```

- The `Time.from_msg` function allows to create a `Time` object from the timestamp of a message.
- The current time is obtained with Node's `get_clock().now()` method.
- The operation between time has as a result an object of type `Duration`, which can be compared with another object of type `Duration`, such as `Duration(seconds = self.BACKING_TIME)` that represents the duration of 2 s.

3.4.3 Running the Code

Let's see how to build and install the code in the workspace. First, Modify `setup.py` for our new program:

setup.py

```
import os
from glob import glob

from setuptools import setup

package_name = 'br2_fsm_bumpgo_py'

setup(
    name=package_name,
    version='0.0.0',
    packages=[package_name],
    data_files=[
        ('share/ament_index/resource_index/packages',
            ['resource/' + package_name]),
        ('share/' + package_name, ['package.xml']),
        (os.path.join('share', package_name, 'launch'), glob('launch/*.launch.py'))
    ],
    install_requires=['setuptools'],
    zip_safe=True,
    maintainer='johndoe',
    maintainer_email='john.doe@evilrobot.com',
    description='BumpGo in Python package',
    license='Apache 2.0',
    tests_require=['pytest'],
    entry_points={
        'console_scripts': [
          'bump_go_main = br2_fsm_bumpgo_py.bump_go_main:main'
        ],
    },
)
```

The important part right now is the `entry_points` argument. As shown in the code above, add the new program shown previously. With this, we can already build our package.

```
$ colcon build --symlink-install
```

In order to run the program, first launch the simuladtor by typing in the terminal:

```
$ ros2 launch br2_tiago sim.launch.py
```

Open another terminal, and run the program:

```
$ ros2 run br2_fsm_bumpgo_py bump_go_main --ros-args -r output_vel:=/nav_vel -r
input_scan:=/scan_raw -p use_sim_time:=true
```

We can also use a launcher similar to the one in the C++ version, just by typing:

```
$ ros2 launch br2_fsm_bumpgo_py bump_and_go.launch.py
```

PROPOSED EXERCISES:

1. Modify the *Bump and Go* project so that the robot perceives an obstacle in front, on its left and right diagonal. Instead of always turning to the same side, it turns to the side with no obstacle.

2. Modify the *Bump and Go* project so that the robot turns exactly to the angle with no obstacles or the more far perceived obstacle. Try two approaches:

   - Open-loop: Calculate before turning time and speed to turn.
   - Closed-loop: Turns until a clear space in front is detected.

CHAPTER 4

The TF Subsystem

ONE of the greatest hidden treasures in ROS is its geometric transformation subsystem TF (or TFs in short). This subsystem allows defining different reference axes (also called frames) and the geometric relationship between them, even when this relationship is constantly changing. Any coordinate in a frame can be recalculated to another frame without the need for tedious manual calculations.

In my experience teaching ROS courses, who has had to deal with similar calculations without TFs, shows a big surprise and happiness when they meet them.

Its importance in ROS is due to the need to model the parts and components of a robot geometrically. It has many applications in navigation and location, as well as manipulation. They have been used to position several cameras in a building or motion capture systems[1].

A robot perceives the environment through sensors placed somewhere on the robot and performs actions for which it needs to specify some spatial position. For instance:

- A distance sensor (laser or RGBD) generates a set of points (x, y, z) that indicate the detected obstacles.
- A robot moves its end effector by specifying a target position $(x, y, z, roll, pitch, yaw)$.
- A robot moves to a point (x, y, yaw) on a map.

All these coordinates are references to a frame. In a robot, there are multiple frames (for sensors, actuators, etc). The relationship between these frames must be known to reason, for example, the coordinate of an obstacle detected by the laser on the arm reference axis to avoid it. Frames relationships are the displacement and rotation of a frame to another frame. Algebraically, this is done using homogeneous coordinates for the coordinates and RT transformation matrices for relations. Having the coordinates of a point P in frame A, this is P_A, we can calculate P_B in frame B using the transformation matrix $RT_{A \to B}$ as follows:

[1]https://github.com/MOCAP4ROS2-Project

DOI: 10.1201/9781003289623-4

$$P_B = RT_{A \to B} * P_A \tag{4.1}$$

$$\begin{pmatrix} x_B \\ y_B \\ z_B \\ 1 \end{pmatrix} = \begin{pmatrix} R^{xx}_{A\to B} & R^{xy}_{A\to B} & R^{xz}_{A\to B} & T^{x}_{A\to B} \\ R^{yx}_{A\to B} & R^{yy}_{A\to B} & R^{yz}_{A\to B} & T^{y}_{A\to B} \\ R^{zx}_{A\to B} & R^{zy}_{A\to B} & R^{zz}_{A\to B} & T^{z}_{A\to B} \\ 0 & 0 & 0 & 1 \end{pmatrix} * \begin{pmatrix} x_A \\ y_A \\ z_A \\ 1 \end{pmatrix} \tag{4.2}$$

In addition to the complexity of these operations, it is remarkable that these relationships are highly dynamic in an articulated robot. It would be an error to transform the points perceived by a sensor at time t using the transformation at t + 0.01 s if it varies dynamically at high speed.

ROS2 implements the TF transform system (now called TF2, the second version) using two topics that receives transformations, as messages of type `tf2_msgs/msg/TFMessage`;

```
$ ros2 interface show tf2_msgs/msg/TFMessage

geometry_msgs/TransformStamped[] transforms
    std_msgs/Header header
    string child_frame_id
    Transform transform
        Vector3 translation
            float64 x
            float64 y
            float64 z
        Quaternion rotation
            float64 x 0
            float64 y 0
            float64 z 0
            float64 w 1
```

- **`/tf`** for transforms that vary dynamically, like the joints of a robot are specified here. By default, they are valid for a short time (10 s). For example, frames relation linked by motorized joints are published here.
- **`/tf_static`** for transforms that do not vary over time. This topic has a QoS transient_local, so any node that subscribes to this topic receives all the transforms published so far. Typically, the transforms published in this topic do not change over time, like the robot geometry.

The frames of a robot are organized as a tree of TFs, in which each TF *should* has at most one parent and can have several children. If this is not true, or several trees are not connected, the robot is not well modeled. By convention, there are several important axes:

- **`/base_footprint`** is usually the root of a robot's TFs, and corresponds to the center of the robot on the ground. It is helpful to transform the information from the robot's sensors to this axis to relate them to each other.

- `/base_link` is usually the child of `/base_footprint`, and is typically the center of the robot, already above ground level.
- `/odom` is the parent frame of `/base_footprint`, and the transformation that relates them indicates the robot's displacement since the robot driver started.

Figure 4.1 shows partially the TF tree of the simulated Tiago. If it is needed to obtain it, launch the simulation and type[2]:

```
$ ros2 run rqt_tf_tree rqt_tf_tree
```

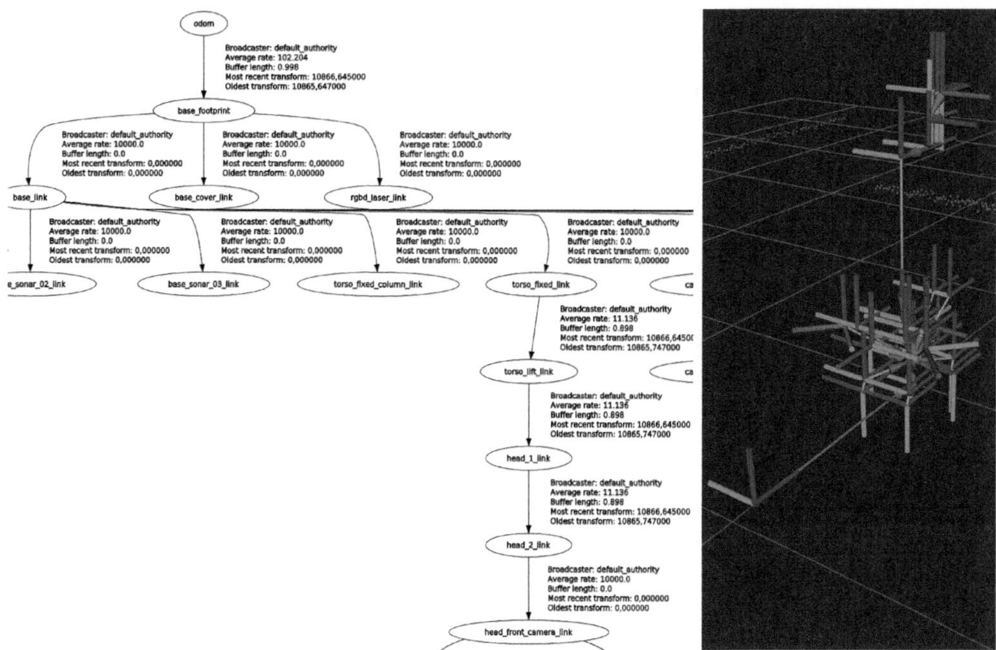

Figure 4.1: Portion of the TF tree of the simulated Tiago and the TF display in RViz2.

When a node wants to use this system, it does not subscribe directly to these topics but uses `TFListeners`, which are objects that update a buffer where are stored all the latest published TFs, and that has an API that lets, for example:

- To know if there is a TF from one frame to another at time t.
- To know what is the rotation from frame A to frame B at time t.
- To ask to transform a coordinate that is in frame A and to frame B in an arbitrary time t.

The buffer may not contain just the TF at time t, but if it has an earlier and a later one, it performs the interpolation. Likewise, frames A and B may not be directly connected, but more frames are in between, performing the necessary matrix operations automatically.

[2]It is needed to have installed the package `ros-foxy-rqt-tf-tree`

Without going into much detail, for now, publishing a transform to a ROS2 node is very straightforward. Just have a transform broadcaster and send transforms to the TF system:

```
geometry_msgs::msg::TransformStamped detection_tf;

detection_tf.header.frame_id = "base_footprint";
detection_tf.header.stamp = now();
detection_tf.child_frame_id = "detected_obstacle";
detection_tf.transform.translation.x = 1.0;

tf_broadcaster_->sendTransform(detection_tf);
```

Getting a transform is easy too. Having a TF buffer that a transform listener updates, we can ask for the geometric transformation from one frame to another. Not even these frames need to be directly connected. Any calculation is done transparently for the developer:

```
tf2_ros::Buffer tfBuffer;
tf2_ros::TransformListener tfListener(tfBuffer);

...

geometry_msgs::msg::TransformStamped odom2obstacle;
odom2obstacle = tfBuffer_.lookupTransform("odom", "detected_obstacle", tf2::TimePointZero);
```

The above code calculates `odom` → `base_footprint` → `detected_obstacle` automatically. The third argument of `lookupTransform` indicates the instant of time from which we want to obtain the transform. `tf2::TimePointZero` indicates the latest available. If we are transforming points of a laser, for example, we should use the timestamp that appears in the header of the laser message, because if a robot or the laser has moved since then, the transformation in another instant will not be exact (much can change in few milliseconds in a robot). Finally, be careful about asking for the transforms with `now()`, because it will not have information yet at this moment in time, and it cannot be extrapolated into the future, and an exception can be raised.

We can operate with transforms, multiplying them or calculating their inverse. From here, we will establish a nomenclature convention in our codes. This will help us to operate with TFs:

- If an object represent a transformation from frame `origin` to frame `target`, we call it `origin2target`.

- If want multiply two TFs, as shown in Figure 4.2.

  1. We only can operate it if the frame names near operator * are equal. In this case, the frame names are equals (robot).
  2. The result frame id must be the outer part of the operators (odom from first operator and object from second).
  3. If we invert a TF (they are invertibles), we invert the frame ids in this name.

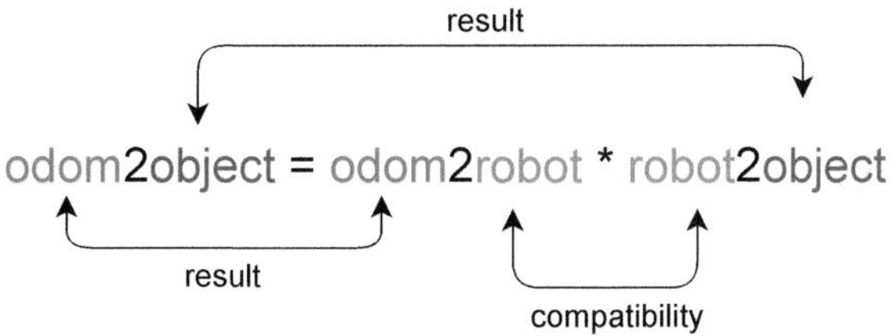

Figure 4.2: The mnemonic rule for naming and operating TFs. Based on their name, we can know if two TFs can be multiplied and the name of the resulting TF.

4.1 AN OBSTACLE DETECTOR THAT USES TF2

This section will analyze a project to see in practice the application of the concepts on TFs set out above.

This project makes the robot detect obstacles right in front of it using the laser sensor, as shown in Figure 4.3.

Figure 4.3: Robot Tiago detecting obstacles with the laser sensor. The red arrow highlights the obstacle detected with the center reading.

We will apply TFs concepts following a common practice in many ROS2 packages to publish the perceptions as TFs. The advantage of doing this is that we can easily reason its position geometrically for any frame, even if it is not currently perceived.

We will not introduce a new **perception model**, but we will use the same one from the previous chapter: we will detect obstacles in front of the robot using the laser. We will use the same speed-based actuation model, although we will teleoperate the robot manually in this case.

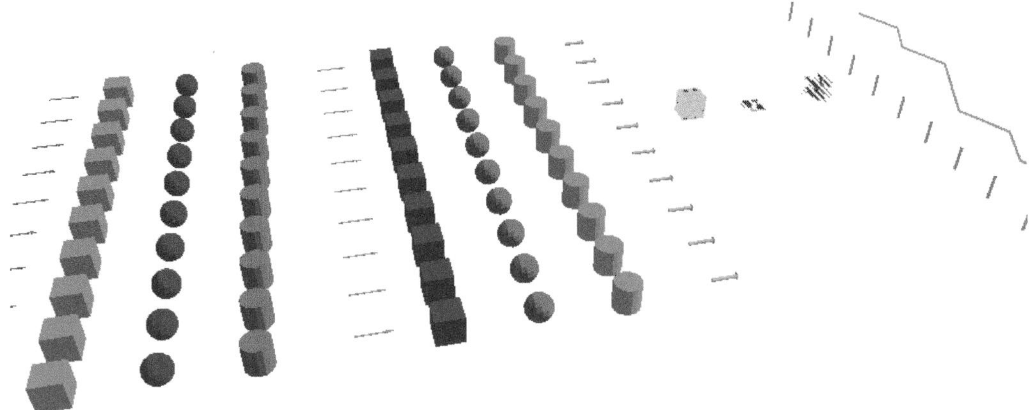

Figure 4.4: Visual markers available for visual debugging.

In this project, apart from using the concepts about TFs, we will show a powerful debugging tool called *Visual Markers*[3], which allows us to publish 3D visual elements that can be viewed in RViz2 from a node. This mechanism allows us to show at a glance part of the internal state of the robot without limiting ourselves to the debugging messages that are generated with the macros `RCLCPP_*`. Markers include arrows, lines, cylinders, spheres, lines, shapes, text, and others in any size or color. Figure 4.4 shows an example of available markers.

4.2 COMPUTATION GRAPH

The Computation Graph (Figure 4.5) of our application is shown in the Figure 4.5.

The node uses a laser sensor of the simulated robot in the `scan_raw` topic. The detection node subscribes to the laser topic and publishes the transform in the ROS2 TF subsystem. Our node subscribes to `/input_scan`, so we will have to remap from `/scan_raw`.

We will create a node `/obstacle_monitor` that reads the transform corresponding to the detection and shows in console its position with respect to the general frame of the robot, `base_footprint`.

The node `/obstacle_monitor` publishes also a visual marker. In our case, we will publish a red arrow that connects the robot's base with the frame's position of the obstacle that we are publishing.

In this project, we will make two versions: a basic one and an improved one. The reason is to see a small detail about the use of TFs that significantly impact the final result, as we will explain later.

4.3 BASIC DETECTOR

We use the same package for both versions. The structure of the package can be seen in the following box:

[3]http://wiki.ros.org/rviz/DisplayTypes/Marker

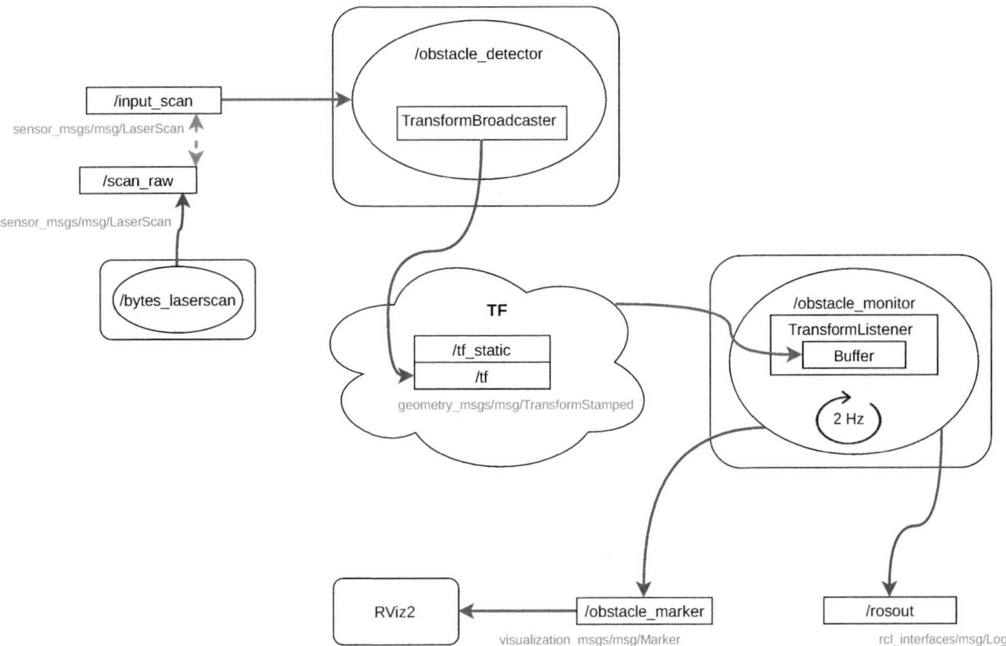

Figure 4.5: Computation Graph of the exercise. The `/obstacle_detector` node collaborates with the `/obstacle_monitor` node using the TF subsystem.

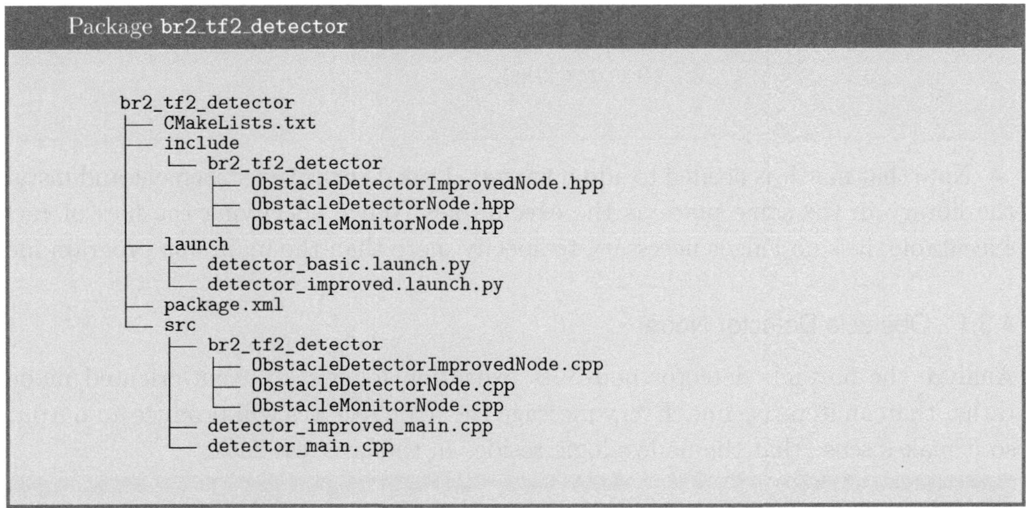
Package `br2_tf2_detector`

```
br2_tf2_detector
├── CMakeLists.txt
├── include
│   └── br2_tf2_detector
│       ├── ObstacleDetectorImprovedNode.hpp
│       ├── ObstacleDetectorNode.hpp
│       └── ObstacleMonitorNode.hpp
├── launch
│   ├── detector_basic.launch.py
│   └── detector_improved.launch.py
├── package.xml
└── src
    ├── br2_tf2_detector
    │   ├── ObstacleDetectorImprovedNode.cpp
    │   ├── ObstacleDetectorNode.cpp
    │   └── ObstacleMonitorNode.cpp
    ├── detector_improved_main.cpp
    └── detector_main.cpp
```

We will ignore in this section the files that contain the word `"Improved"` in the name. We will see them in the next section.

The reader can see how the package structure is similar to the previous chapter. The nodes are separated in their declaration and definition, in directories whose name matches the package. In addition, everything will be defined within a namespace that matches the package's name. This package will take a small step forward in this structure: now, we will compile the nodes as a dynamic library linked by the executables. Perhaps in this project we will not notice the difference, but we save space, it is more convenient, and it could allow (it is not the case) to export it to other packages. The name of the library will be the name of the package (`${PROJECT_NAME}`), as usual when creating a support library in a package. Let's see what this looks like in the `CMakeLists.txt` file:

include/br2_tf2_detector/ObstacleDetectorNode.hpp

```
project(br2_tf2_detector)

find_package(...)
...

set(dependencies
  ...
)

include_directories(include)

add_library(${PROJECT_NAME} SHARED
  src/br2_tf2_detector/ObstacleDetectorNode.cpp
  src/br2_tf2_detector/ObstacleMonitorNode.cpp
  src/br2_tf2_detector/ObstacleDetectorImprovedNode.cpp
)
ament_target_dependencies(${PROJECT_NAME} ${dependencies})

add_executable(detector src/detector_main.cpp)
ament_target_dependencies(detector ${dependencies})
target_link_libraries(detector ${PROJECT_NAME})

add_executable(detector_improved src/detector_improved_main.cpp)
ament_target_dependencies(detector_improved ${dependencies})
target_link_libraries(detector_improved ${PROJECT_NAME})

install(TARGETS
  ${PROJECT_NAME}
  detector
  detector_improved
  ARCHIVE DESTINATION lib
  LIBRARY DESTINATION lib
  RUNTIME DESTINATION lib/${PROJECT_NAME}
)
```

Note that now it is needed to add a `target_link_libraries` statement and install the library in the same place as the executables. When specifying the files of each executable, it is no longer necessary to specify more than the main `cpp` program file.

4.3.1 Obstacle Detector Node

Analyze the obstacle detector node. Its execution follows an event-oriented model rather than an iterative one. Every message the node receives will produce an output, so it makes sense that the node's logic resides in the laser callback.

include/br2_tf2_detector/ObstacleDetectorNode.hpp

```
class ObstacleDetectorNode : public rclcpp::Node
{
public:
  ObstacleDetectorNode();

private:
  void scan_callback(sensor_msgs::msg::LaserScan::UniquePtr msg);

  rclcpp::Subscription<sensor_msgs::msg::LaserScan>::SharedPtr scan_sub_;
  std::shared_ptr<tf2_ros::StaticTransformBroadcaster> tf_broadcaster_;
};
```

Since the node must publish transforms to the TF subsystem, we declare a `StaticTransformBroadcaster`, that publish in `/tf_static`. We could also declare a `TransformBroadcaster`, that publish in `/tf`. Apart from the durability QoS, the difference is that we want transforms to persist beyond the 10 s by default of non-static transforms.

We use a `shared_ptr` for `tf_broadcaster_`, since its constructor requires an `rclcpp::Node*`, and we will not have it until we are already inside the constructor[4]:

src/br2_tf2_detector/ObstacleDetectorNode.hpp

```
ObstacleDetectorNode::ObstacleDetectorNode()
: Node("obstacle_detector")
{
  scan_sub_ = create_subscription<sensor_msgs::msg::LaserScan>(
    "input_scan", rclcpp::SensorDataQoS(),
    std::bind(&ObstacleDetectorNode::scan_callback, this, _1));

  tf_broadcaster_ = std::make_shared<tf2_ros::TransformBroadcaster>(*this);
}
```

The `tf_broadcaster_` object manages the publication of static TFs. The message type of a TF is `geometry_msgs/msg/TransformStamped`. Let's see how it is used:

src/br2_tf2_detector/ObstacleDetectorNode.hpp

```
void
ObstacleDetectorNode::scan_callback(sensor_msgs::msg::LaserScan::UniquePtr msg)
{
  double dist = msg->ranges[msg->ranges.size() / 2];

  if (!std::isinf(dist)) {
    geometry_msgs::msg::TransformStamped detection_tf;

    detection_tf.header = msg->header;
    detection_tf.child_frame_id = "detected_obstacle";
    detection_tf.transform.translation.x = msg->ranges[msg->ranges.size() / 2];

    tf_broadcaster_->sendTransform(detection_tf);
  }
}
```

- The header of the output message will be the header of the input laser message. We will do this because the timestamp must be when the sensory reading was taken. If we used `now()`, depending on the latency in the messages and the load of the computer, the transform would not be precise, and synchronization errors could occur.

  The `frame_id` is the source frame (or parent frame) of the transformation, already in this header. In this case, it is the sensor frame since the perceived coordinates of the object are in this frame.

- The `child_frame_id` field is the id of the new frame that we are going to create, and that represents the perceived obstacle.

[4]In fact, some C++ developers recommend avoiding using `this` in constructors, as the object has not completely initialized until the constructor finishes.

- The transform field contains a translation and a rotation applied in this order, from the parent frame to the child frame that we want to create. Since the X-axis of the laser frame is aligned with the laser beam that we are measuring, the translation in X is the distance read.

  Rotation refers to the rotation of the frame after translation is applied. As this value is not relevant here (detection is a point) we use the default quaternion values $(0, 0, 0, 1)$ set by the message constructor.

- Finally, use the `sendTransform()` method of `tf_broadcaster_` to send the transform to the TF subsystem.

4.3.2 Obstacle Monitor Node

The `/obstacle_monitor` node extracts the transform to the detected object from the TFs system and shows it to the user in two ways:

- The standard output on the console indicates where the obstacle is with respect to the robot at all times, even if it is no longer being detected.

- Using a visual marker, specifically an arrow, which starts from the robot toward the obstacle that was detected.

Analyze the header to see what elements this node has:

include/br2_tf2_detector/ObstacleMonitorNode.hpp

```
class ObstacleMonitorNode : public rclcpp::Node
{
public:
  ObstacleMonitorNode();

private:
  void control_cycle();
  rclcpp::TimerBase::SharedPtr timer_;

  tf2::BufferCore tf_buffer_;
  tf2_ros::TransformListener tf_listener_;

  rclcpp::Publisher<visualization_msgs::msg::Marker>::SharedPtr marker_pub_;
};
```

- The execution model of this node is iterative, so we declare `timer_` and its callback `control_cycle`.

- To access the TF system, use a `tf2_ros::TransformListener` that update the buffer `tf_buffer_` to which we can make the queries we need.

- We only need one publisher for visual markers.

In the case of the class definition, we ignore the part dedicated to visual markers, for now, showing only the part related to TFs.

src/br2_tf2_detector/ObstacleMonitorNode.cpp

```
ObstacleMonitorNode::ObstacleMonitorNode()
: Node("obstacle_monitor"),
  tf_buffer_(),
  tf_listener_(tf_buffer_)
{
  marker_pub_ = create_publisher<visualization_msgs::msg::Marker>(
    "obstacle_marker", 1);

  timer_ = create_wall_timer(
    500ms, std::bind(&ObstacleMonitorNode::control_cycle, this));
}

void
ObstacleMonitorNode::control_cycle()
{
  geometry_msgs::msg::TransformStamped robot2obstacle;

  try {
    robot2obstacle = tf_buffer_.lookupTransform(
      "base_footprint", "detected_obstacle", tf2::TimePointZero);
  } catch (tf2::TransformException & ex) {
    RCLCPP_WARN(get_logger(), "Obstacle transform not found: %s", ex.what());
    return;
  }

  double x = robot2obstacle.transform.translation.x;
  double y = robot2obstacle.transform.translation.y;
  double z = robot2obstacle.transform.translation.z;
  double theta = atan2(y, x);

  RCLCPP_INFO(get_logger(), "Obstacle detected at (%lf m, %lf m, , %lf m) = %lf rads",
    x, y, z, theta);
}
```

- Notice how `tf_listener_` is initialized, simply specifying the buffer to update. Later, the queries will be made directly to the buffer.
- We observe that the control loop runs at 2 Hz, showing us information with `RCLCPP_INFO` (to `/ros_out` and stdout).
- The most relevant function is `lookupTransform`, which calculates the geometric transformation from one frame to another, even if there is no direct relationship. We can specify a specific timestamp or, on the contrary, we want the last one available by indicating `tf2::TimePointZero`. This call can throw an exception if it does not exist, or we require a transform on a timestamp in the future, so a `try/catch` should be used to handle possible errors.
- Note that the TF we published in `ObstacleDetectorNode` was `base_laser_link` → `detected_obstacle`, and now we are requiring `base_footprint` → `detected_obstacle`. As the robot is well modeled and the geometric relationship between `base_laser_link` and `base_footprint` can be calculated, there will be no problem for `lookupTransform` to return the correct information.

Let's see the part related to the generation of the visual marker. The goal is to show the coordinates of the obstacle to the robot on the screen and show a geometric shape in RViz2 that allows us to debug the application visually. In this case, it will be a red arrow from the robot to the obstacle. To do this, create an `visualization_msgs/msg/Marker` message and fill in its fields to obtain this arrow:

src/br2_tf2_detector/ObstacleMonitorNode.cpp

```
visualization_msgs::msg::Marker obstacle_arrow;
obstacle_arrow.header.frame_id = "base_footprint";
obstacle_arrow.header.stamp = now();
obstacle_arrow.type = visualization_msgs::msg::Marker::ARROW;
obstacle_arrow.action = visualization_msgs::msg::Marker::ADD;
obstacle_arrow.lifetime = rclcpp::Duration(1s);

geometry_msgs::msg::Point start;
start.x = 0.0;
start.y = 0.0;
start.z = 0.0;
geometry_msgs::msg::Point end;
end.x = x;
end.y = y;
end.z = z;
obstacle_arrow.points = {start, end};

obstacle_arrow.color.r = 1.0;
obstacle_arrow.color.g = 0.0;
obstacle_arrow.color.b = 0.0;
obstacle_arrow.color.a = 1.0;

obstacle_arrow.scale.x = 0.02;
obstacle_arrow.scale.y = 0.1;
obstacle_arrow.scale.z = 0.1;
```

In the reference document[5] for visual markers is documented the meaning of every field for every type of marker. In the case of an arrow, the points field will be filled with the starting point $(0, 0, 0)$ and the ending point corresponding to the detection, both in `base_footprint`. Do not forget to assign a color, especially the `alpha`, since we will not see anything if we let it be 0, the default value.

4.3.3 Running the Basic Detector

We instantiate both in the same process to test our nodes, and we use a `SingleThreadedExecutor`. That would be enough to spin both:

src/br2_tf2_detector/src/detector_main.cpp

```
int main(int argc, char * argv[]) {
  rclcpp::init(argc, argv);

  auto obstacle_detector = std::make_shared<br2_tf2_detector::ObstacleDetectorNode>();
  auto obstacle_monitor = std::make_shared<br2_tf2_detector::ObstacleMonitorNode>();

  rclcpp::executors::SingleThreadedExecutor executor;
  executor.add_node(obstacle_detector->get_node_base_interface());
  executor.add_node(obstacle_monitor->get_node_base_interface());

  executor.spin();

  rclcpp::shutdown();
  return 0;
}
```

Follow the next commands to test our nodes:

```
# Terminal 1: The Tiago simulation
$ ros2 launch br2_tiago sim.launch.py world:=empty
```

[5] http://wiki.ros.org/rviz/DisplayTypes/Marker

```
# Terminal 2: Launch our nodes
$ ros2 launch br2_tf2_detector detector_basic.launch.py
```

```
# Terminal 3: Keyboard teleoperation
$ ros2 run teleop_twist_keyboard teleop_twist_keyboard --ros-args -r
cmd_vel:=/key_vel
```

```
# Terminal 4: RViz2
$ ros2 run rviz2 rviz2 --ros-args -p use_sim_time:=true
```

In Gazebo, add an obstacle in front of the robot. Start watching in the terminal the information about the detection. In Rviz2, change the fixed frame to `odom`. Add a Markers display to RViz2 specifying the topic that we have created to publish the visual marker. Also, add the TF Display if it is not added yet. Figure 4.6 shows the TF to the obstacle and also the red arrow.

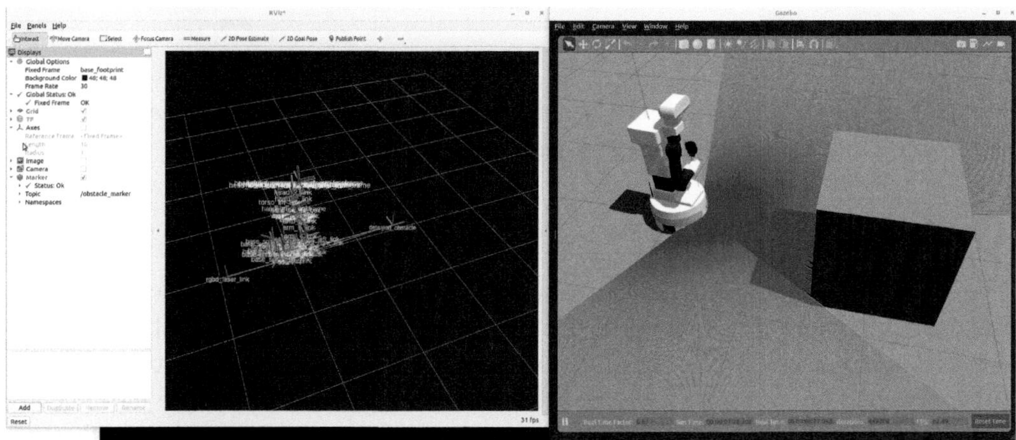

Figure 4.6: Visualization in RViz2 of the TF corresponding to the detection, and the red arrow marker published to visualize the detection.

Do a quick exercise: with the teleoperator, move the robot forward and to the side, so it no longer perceives the obstacle. Keep moving the robot and realize that the information returned by `lookupTransform` is no longer correct. It continues to indicate that the obstacle is ahead, although this is no longer true. What has happened? We probably wanted the arrow to point to the obstacle position, but now the arrow travels fixed with the robot.

Let's explain it with a diagram in Figure 4.7. As long as the robot perceives the obstacle, the transform requested (pink arrow) is correct. It is a transform from the robot's laser to the obstacle. When we stop updating the transform (thick blue arrow) because the obstacle is gone, the transform continues to exist. If we move the robot, `lookupTransform` keeps returning the last valid transform: in front fo the robot. This makes the visual marker wrong as well. The following section presents a strategy to fix this undesirable situation.

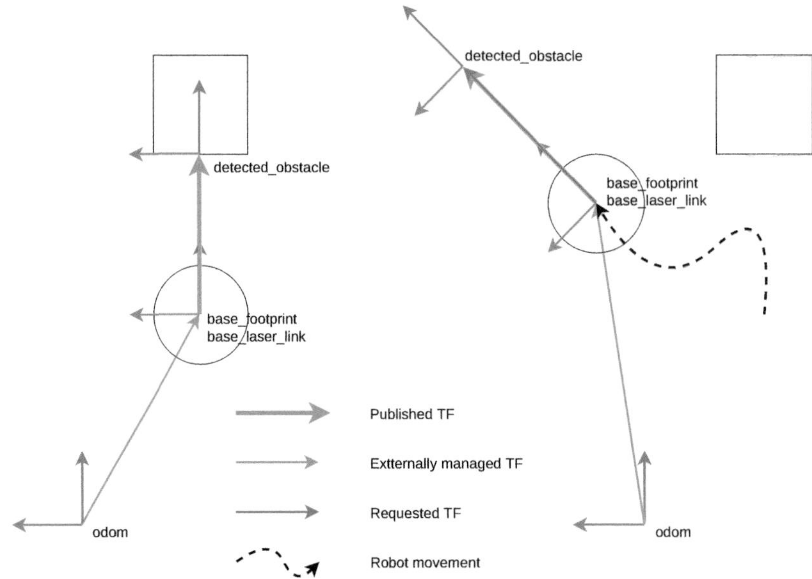

Figure 4.7: Diagram showing the problem when publishing TFs in the local frame. When the robot moves, the TF no longer represents the right obstacle position.

4.4 IMPROVED DETECTOR

The solution is to publish the detection TF in a fixed frame that is not affected by the robot's movement, for example, `odom` (or `map` if your robot is navigating). If we do it like this, when we require the transform `base_footprint` → `detected_obstacle` (pink arrow), this transform will be calculated taking into account the movement of the robot, collected in the transformation `odom` → `base_footprint`. It is shown in the diagram in Figure 4.8.

`ObstacleDetectorImprovedNode` is the modification of `ObstacleDetectorNode` to implement this improvement. This new node operates with TFs, so at some point, it consult the value of an existing TF. For this reason, in addition to having a `StaticTransformPublisher`, it instantiates a `TransformListener` with its related `Buffer`.

include/br2_tf2_detector/ObstacleMonitorNode.hpp

```
class ObstacleDetectorImprovedNode : public rclcpp::Node
{
...
private:
  ...
  tf2::BufferCore tf_buffer_;
  tf2_ros::TransformListener tf_listener_;
};
```

Let the implementation of this node. In this program, check the two data structures that are related but are not the same:

- **`geometry_msgs::msg::TransformStamped`** is a message type, and is used to post TFs, and is the returned result of `lookupTransform`.

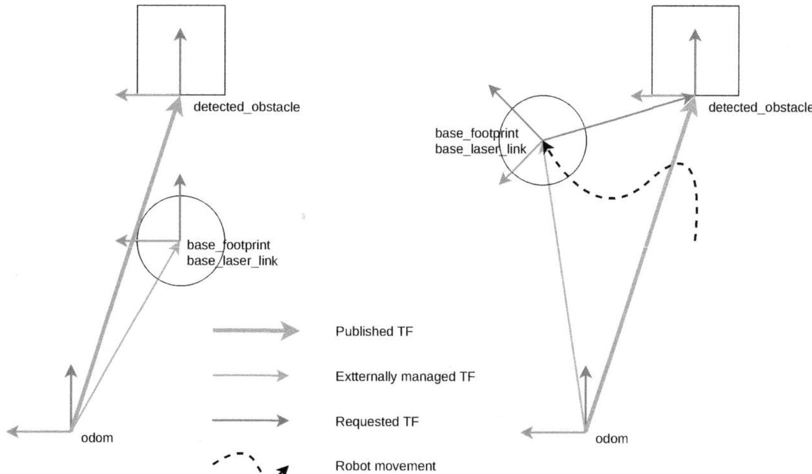

Figure 4.8: Diagram showing how to correctly maintain the obstacle position, by publishing the TF in a fixed frame. The calculated TF (thick blue arrow) takes into account the robot displacement.

- **`tf2::Transform`** It is a data type of the TF2 library that allows to perform operations.
- **`tf2::Stamped<tf2::Transform>`** is similar to the previous one, but with a header that indicates a timestamp. It will be necessary to comply with the types in the transformation functions.
- **`tf2::fromMsg/tf2::toMsg`** are transformation functions that allow transforming from a message type to a TF2 type, and vice versa.

As a general tip, do not use message types inside the node to operate on them. Apply this advice for TFs, images, point clouds, and more data type. Messages are good for communicating nodes but very limited in functionality. If there is a library that offers a native type, use it, as it will be much more useful. Commonly, there are functions to pass from message type to native type. In this case, we use `geometry_msgs::msg::TransformStamped` to send and receive TFs, but we use the TF2 library to operate on them.

Considering the convention established previously, let's see how we can carry out the improvement. Our goal, as we saw before, is to create the TF `odom2object` (object is the detected obstacle). The observation is represented as the transform `laser2object`, so we have to find X in the following equation:

$$odom2object = X * laser2object$$

By deduction from the rules that we stated above, X must be `odom2laser`, which is a TF that can be requested from `lookupTransform`.

include/br2_tf2_detector/ObstacleMonitorNode.hpp

```
double dist = msg->ranges[msg->ranges.size() / 2];

if (!std::isinf(dist)) {
  tf2::Transform laser2object;
  laser2object.setOrigin(tf2::Vector3(dist, 0.0, 0.0));
  laser2object.setRotation(tf2::Quaternion(0.0, 0.0, 0.0, 1.0));

  geometry_msgs::msg::TransformStamped odom2laser_msg;
  tf2::Stamped<tf2::Transform> odom2laser;
  try {
    odom2laser_msg = tf_buffer_.lookupTransform(
      "odom", "base_laser_link", msg->header.stamp, rclcpp::Duration(200ms));
    tf2::fromMsg(odom2laser_msg, odom2laser);
  } catch (tf2::TransformException & ex) {
    RCLCPP_WARN(get_logger(), "Obstacle transform not found: %s", ex.what());
    return;
  }

  tf2::Transform odom2object = odom2laser * laser2object;

  geometry_msgs::msg::TransformStamped odom2object_msg;
  odom2object_msg.transform = tf2::toMsg(odom2object);

  odom2object_msg.header.stamp = msg->header.stamp;
  odom2object_msg.header.frame_id = "odom";
  odom2object_msg.child_frame_id = "detected_obstacle";

  tf_broadcaster_->sendTransform(odom2object_msg);
}
```

- `laser2object` stores the perception to the detected object. It is just a translation in the X-axis corresponding to the distance to the obstacle.
- To get `odom2laser`, we need to use query the TF subsystem with `lookupTransform`, transforming the resulting transform message to the needed type to operate with transforms.
- At this point, we have everything to calculate `odom2object`, this is, the obstacle position with respect to the fixed frame `odom`.
- Finally, we compound the output message and publish to the TF subsystem.

It is not necessary to make any changes to the `ObstacleMonitorNode` as `lookupTransform` will calculate the TF `base_footprint` → `obstacle` since the TF system knows the TFs `odom` → `base_footprint` and `odom` → `obstacle`.

4.4.1 Running the Improved Detector

The process of executing the nodes is similar to the basic case. In this case, the main program and a launcher is similar to the one in the basic case, but with the new improved node, so we will skip showing it here. Let's follow similar commands to execute it:

```
# Terminal 1: The Tiago simulation
$ ros2 launch br2_tiago sim.launch.py world:=empty
```

```
# Terminal 2: Launch our nodes
$ ros2 launch br2_tf2_detector detector_improved.launch.py
```

```
# Terminal 3: Keyboard teleoperation
$ ros2 run teleop_twist_keyboard teleop_twist_keyboard --ros-args -r
cmd_vel:=/key_vel
```

```
# Terminal 4: RViz2
$ ros2 run rviz2 rviz2 --ros-args -p use_sim_time:=true
```

Add the obstacle in Gazebo so the robot can detect it. Watch the console output and the visual marker in RViz2. Move the robot so the obstacle is not detected, and see how the marker and the output are correct now. The displacement, coded as the transform `odom` → `base_footprint` is used to update the information correctly.

PROPOSED EXERCISES:

1. Make a node that shows every second how much the robot has moved. You can do this by saving $(odom \rightarrow base\_footprint)_t$, and subtracting it from $(odom \rightarrow base\_footprint)_{t+1}$

2. In `ObstacleDetectorNode`, change the arrow's color depending on the distance to the obstacle: green is far, and red is near.

3. In `ObstacleDetectorNode`, show in the terminal the obstacle's position in the `odom` frame, in `base_footprint`, and `head_2_link`.

CHAPTER 5

Reactive Behaviors

REACTIVE behaviors tightly couples perception to action without the use of intervening abstract representation. As Brooks demonstrated in his Subsumption Architectures[1], relatively complex behaviors can be created with simple reactive behaviors that are activated or inhibited by higher layers.

We will not discuss the development of sumbsumption architectures in this chapter. By the way, the reader can refer to the Cascade Lifecycle[1] package and rqt_cascade_hfsm[2], which provide some building blocks to build subsumption architectures. The objective of this chapter is to show a couple of reactive behaviors that use different resources to advance the knowledge of ROS2.

This chapter will first look at a simple local navigation algorithm, Virtual Force Field (VFF), that uses the laser to avoid obstacles. This example will establish some knowledge about visual markers and introduce some test-driven development methodology.

Second, we will see reactive tracking behavior based on information from the camera. We will see how images are processed and how the joints of a robot are controlled. In addition, we will see an advantageous type of node called Lifecycle Node.

5.1 AVOIDING OBSTACLES WITH VFF

This section will show how to implement a simple reactive behavior that makes the Tiago robot move forward, avoiding obstacles using a simple VFF algorithm. This simple algorithm is based on using three 2D vectors to calculate the control speed:

- **Attractive vector**: This vector always points forward since the robot wants to move in a straight line in the absence of obstacles.
- **Repulsive vector**: This vector is calculated from the laser sensor readings. In our basic version, the obstacle closest to the robot produces a repulsion vector, inversely proportional to its distance.

[1] https://github.com/fmrico/cascade_lifecycle
[2] https://github.com/fmrico/rqt_cascade_hfsm

DOI: 10.1201/9781003289623-5

- **Result vector**: This vector is the sum of the two previous vectors and will calculate the control speed. Linear speed depends on the resulting vector module, and the angle to turn depends on the resulting vector angle.

Figure 5.1 shows examples of these vectors depending on the position of the obstacles.

Figure 5.1: Examples of VFF vectors produced by the same obstacle. Blue vector is attractive, red vector is repulsive, and green vector is the resulting vector.

5.1.1 The Computation Graph

First, see what the computational graph of this problem looks like. As shown in Figure 5.2, we have a single node within a process, with the following elements and characteristics:

- The node subscribes to a message topic with the perception information and publishes it to a speed message topic. These will be the main input and outputs. As discussed in the previous chapter, we will use generic names for these topics, which will be remapped at deployment.

- It is crucial to have enough information to determine why a robot behaves in a certain way. ROS2 offers many debugging tools. Using `/rosout` is a good alternative. It is also handy to use the LEDs equipped by a robot. With an LED that could change color, we could already color-code the robot's state or perception. At a glance, we could have much information about why the robot makes its decisions.

  In this case, in addition to the input and output topics above, we have added the debugging topic `/vff_debug`, that publish Visual Markers to visualize the

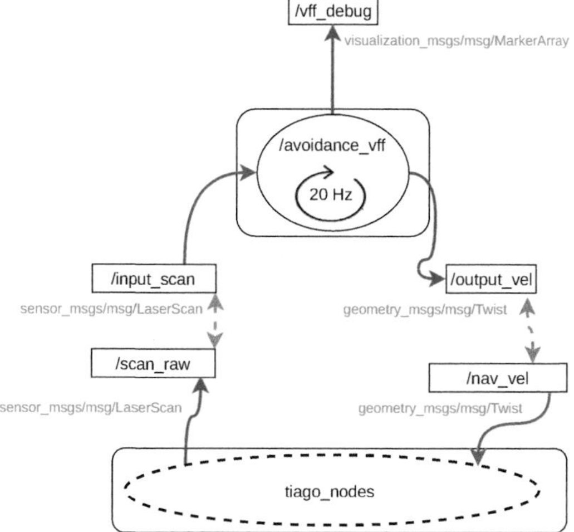

Figure 5.2: Computation Graph for obstacle avoidance.

different vectors of VFF. The color vectors in Figure 5.1 are visual markers published by the node and visualized in RViz2.

- In this case, we will choose an iterative execution controlled internally by the node using a timer, to run the control logic at 20 Hz.

5.1.2 Package Structure

See that the organization the package, in the next box, is already standard in our packages: Each node with its declaration and its different definition in its `.hpp` and its `.cpp`, and the main program that will instantiate it. We have a launch directory with a launcher to easily execute our project. Notice that we have now added a tests directory in which we will have our files with the tests, as we will explain later.

Package br2_vff_avoidance

```
br2_vff_avoidance
├── CMakeLists.txt
├── include
│   └── br2_vff_avoidance
│       └── AvoidanceNode.hpp
├── launch
│   └── avoidance_vff.launch.py
├── package.xml
├── src
│   ├── avoidance_vff_main.cpp
│   └── br2_vff_avoidance
│       └── AvoidanceNode.cpp
└── tests
    ├── CMakeLists.txt
    └── vff_test.cpp
```

5.1.3 Control Logic

The AvoidanceNode implements the VFF algorithm to generate the control commands based on the laser readings. The main elements are similar to the previous examples:

- A subscriber for the laser readings, whose function will be to update the last reading in last_scan_.
- A publisher for speeds.
- A get_vff function for calculating the three vectors on which the VFF algorithm is based, given a reading from the laser. We declare a new type VFFVectors to pack them.
- As this node executes iteratively, we use a timer and use the method control_cycle as a callback.

include/br2_vff_avoidance/AvoidanceNode.hpp

```
struct VFFVectors
{
  std::vector<float> attractive;
  std::vector<float> repulsive;
  std::vector<float> result;
};

class AvoidanceNode : public rclcpp::Node
{
public:
  AvoidanceNode();

  void scan_callback(sensor_msgs::msg::LaserScan::UniquePtr msg);
  void control_cycle();

protected:
  VFFVectors get_vff(const sensor_msgs::msg::LaserScan & scan);

private:
  rclcpp::Publisher<geometry_msgs::msg::Twist>::SharedPtr vel_pub_;
  rclcpp::Subscription<sensor_msgs::msg::LaserScan>::SharedPtr scan_sub_;
  rclcpp::TimerBase::SharedPtr timer_;

  sensor_msgs::msg::LaserScan::UniquePtr last_scan_;
};
```

In the control cycle, initially check if the laser has new data. If not, or if this data is old (if we have not received information from the laser in the last second), do not generate control commands. The robot should stop if the robot driver is correctly implemented and does not move when it stops receiving commands. On the contrary (not our case), you should send speeds with all fields to 0 to stop the robot.

Once the resulting vector has been calculated, its transformation at speeds is direct by calculating modulus and angle. It is convenient to control that the speed ranges are in safe ranges with std::clamp, as can be seen in the following code:

src/br2_vff_avoidance/AvoidanceNode.cpp

```
void
AvoidanceNode::scan_callback(sensor_msgs::msg::LaserScan::UniquePtr msg)
{
  last_scan_ = std::move(msg);
}

void
AvoidanceNode::control_cycle()
{
  // Skip cycle if no valid recent scan available
  if (last_scan_ == nullptr || (now() - last_scan_->header.stamp) > 1s) {
    return;
  }

  // Get VFF vectors
  const VFFVectors & vff = get_vff(*last_scan_);

  // Use result vector to calculate output speed
  const auto & v = vff.result;
  double angle = atan2(v[1], v[0]);
  double module = sqrt(v[0] * v[0] + v[1] * v[1]);

  // Create ouput message, controlling speed limits
  geometry_msgs::msg::Twist vel;
  vel.linear.x = std::clamp(module, 0.0, 0.3);  // linear vel to [0.0, 0.3] m/s
  vel.angular.z = std::clamp(angle, -0.5, 0.5);  // rotation vel to [-0.5, 0.5] rad/s

  vel_pub_->publish(vel);
}
```

5.1.4 Calculation of the VFF Vectors

The objective of the function `get_vff` is to obtain the three vectors: attractive, repulsive, and resulting:

src/br2_vff_avoidance/AvoidanceNode.cpp

```
VFFVectors
AvoidanceNode::get_vff(const sensor_msgs::msg::LaserScan & scan)
{
  // This is the obstacle radious in which an obstacle affects the robot
  const float OBSTACLE_DISTANCE = 1.0;

  // Init vectors
  VFFVectors vff_vector;
  vff_vector.attractive = {OBSTACLE_DISTANCE, 0.0};  // Robot wants to go forward
  vff_vector.repulsive = {0.0, 0.0};
  vff_vector.result = {1.0, 0.0};

  // Get the index of nearest obstacle
  int min_idx = std::min_element(scan.ranges.begin(), scan.ranges.end())
    - scan.ranges.begin();

  // Get the distance to nearest obstacle
  float distance_min = scan.ranges[min_idx];

  // If the obstacle is in the area that affects the robot, calculate repulsive vector
  if (distance_min < OBSTACLE_DISTANCE) {
    float angle = scan.angle_min + scan.angle_increment * min_idx;

    float oposite_angle = angle + M_PI;
    // The module of the vector is inverse to the distance to the obstacle
    float complementary_dist = OBSTACLE_DISTANCE - distance_min;

    // Get cartesian (x, y) components from polar (angle, distance)
    vff_vector.repulsive[0] = cos(oposite_angle) * complementary_dist;
    vff_vector.repulsive[1] = sin(oposite_angle) * complementary_dist;
  }
```

src/br2_vff_avoidance/AvoidanceNode.cpp

```
  // Calculate resulting vector adding attractive and repulsive vectors
  vff_vector.result[0] = (vff_vector.repulsive[0] + vff_vector.attractive[0]);
  vff_vector.result[1] = (vff_vector.repulsive[1] + vff_vector.attractive[1]);

  return vff_vector;
}
```

- The attractive vector will always be $(1, 0)$, since the robot will always try to move forward. Initialize the rest of the vectors assuming there are no nearby obstacles.
- The repulsive vector is calculated from the lower laser reading. By calculating `min_idx` as the index of the vector with a smaller value, we are able to get the distance (the value in the ranges vector) and the angle (from `angle_min`, the `angle_increment` and the `min_idx`).
- The margnitude of the repulsive vector has to be inversely proportional to the distance to the obstacle. Closer obstacles have to generate more repulse than those close.
- The angle of the repulsive vector must be in the opposite direction to the angle of the detected obstacle, so add π to it.
- After calculating the repulsive vector's cartesian coordinates, we add it with the attractive vector to obtain its resultant.

5.1.5 Debugging with Visual Markers

In the previous chapter we used visual markers to visually debug the robot's behavior. The arrows in Figure 5.1 are visual markers generated by `AvoidanceNode` for debugging. The difference is using `visualization_msgs::msg::MarkerArray` instead of `visualization_msgs::msg::Marker`. Basically, a `visualization_msgs::msg::MarkerArray` contains a `std::vector` of `visualization_msgs::msg::Marker` in its field `markers`. Let's see how the message that will be published as debugging information is composed. For details of these messages check the message definitions, and the reference page[3]:

```
$ ros2 interface show visualization_msgs/msg/MarkerArray
Marker[] markers

$ ros2 interface show visualization_msgs/msg/Marker
```

The `AvoidanceNode` header contains what you need to compose and publish the visual markers. We have a publisher of `visualization_msgs::msg::MarkerArray` and two functions that will help us to compose the vectors. `get_debug_vff` returns the complete message formed by the three arrows that represent the three vectors.

[3] http://wiki.ros.org/rviz/DisplayTypes/Marker

To avoid repeating code in this function, make_marker creates a marker with the specified color as the input parameter.

include/br2_vff_avoidance/AvoidanceNode.hpp

```
typedef enum {RED, GREEN, BLUE, NUM_COLORS} VFFColor;

class AvoidanceNode : public rclcpp::Node
{
public:
  AvoidanceNode();

protected:
  visualization_msgs::msg::MarkerArray get_debug_vff(const VFFVectors & vff_vectors);
  visualization_msgs::msg::Marker make_marker(
    const std::vector<float> & vector, VFFColor vff_color);

private:
  rclcpp::Publisher<visualization_msgs::msg::MarkerArray>::SharedPtr vff_debug_pub_;
};
```

The markers are published in control_cycle, as long as there is a subscriber interested in this information, which, in this case, will be RViz2.

```
void
AvoidanceNode::control_cycle()
{
  // Get VFF vectors
  const VFFVectors & vff = get_vff(*last_scan_);

  // Produce debug information, if any interested
  if (vff_debug_pub_->get_subscription_count() > 0) {
    vff_debug_pub_->publish(get_debug_vff(vff));
  }
}
```

For each of the vectors, create a visualization_msgs::msg::Marker with a different color. base_fooprint is the frame that is on the ground, in the center of the robot, facing forward. So, the arrow's origin is (0, 0) in this frame, and the arrow's end is what each vector indicates. Each vector must have a different id since a marker will replace another with the same id in RViz2.

```
visualization_msgs::msg::MarkerArray
AvoidanceNode::get_debug_vff(const VFFVectors & vff_vectors)
{
visualization_msgs::msg::MarkerArray marker_array;

marker_array.markers.push_back(make_marker(vff_vectors.attractive, BLUE));
marker_array.markers.push_back(make_marker(vff_vectors.repulsive, RED));
marker_array.markers.push_back(make_marker(vff_vectors.result, GREEN));

return marker_array;
}

visualization_msgs::msg::Marker
AvoidanceNode::make_marker(const std::vector<float> & vector, VFFColor vff_color)
{
  visualization_msgs::msg::Marker marker;

  marker.header.frame_id = "base_footprint";
  marker.header.stamp = now();
  marker.type = visualization_msgs::msg::Marker::ARROW;
  marker.id = visualization_msgs::msg::Marker::ADD;

  geometry_msgs::msg::Point start;
  start.x = 0.0;
```

```
  start.y = 0.0;
  geometry_msgs::msg::Point end;
  start.x = vector[0];
  start.y = vector[1];
  marker.points = {end, start};

  marker.scale.x = 0.05;
  marker.scale.y = 0.1;

  switch (vff_color) {
    case RED:
      marker.id = 0;
      marker.color.r = 1.0;
      break;
    case GREEN:
      marker.id = 1;
      marker.color.g = 1.0;
      break;
    case BLUE:
      marker.id = 2;
      marker.color.b = 1.0;
      break;
  }
  marker.color.a = 1.0;

  return marker;
}
```

5.1.6 Running the AvoidanceNode

The main program that runs this node should now be pretty trivial to the reader. Just instantiate the node and call with it to spin:

src/avoidance_vff_main.cpp

```
int main(int argc, char * argv[])
{
  rclcpp::init(argc, argv);

  auto avoidance_node = std::make_shared<br2_reactive_behaviors::AvoidanceNode>();
  rclcpp::spin(avoidance_node);

  rclcpp::shutdown();

  return 0;
}
```

To run this node, we must first run the simulator:

```
$ ros2 launch mr2_tiago sim.launch.py
```

Next, execute the node setting remaps and parameters:

```
$ ros2 run br2_vff_avoidance avoidance_vff --ros-args -r input_scan:=/scan_raw -r
output_vel:=/key_vel -p use_sim_time:=true
```

Or using the launcher:

```
$ ros2 launch br2_vff_avoidance avoidance_vff.launch.py
```

If everything goes well, the robot starts to move forward. Use the buttons to move objects in the simulator to put obstacles to the robot. Open RViz2 and add the visualization of topic /vff_debug of type visualization_msgs::msg::MarkerArray, as shown in Figure 5.3. See how the visual information of the node's markers helps us better understand what the robot is doing.

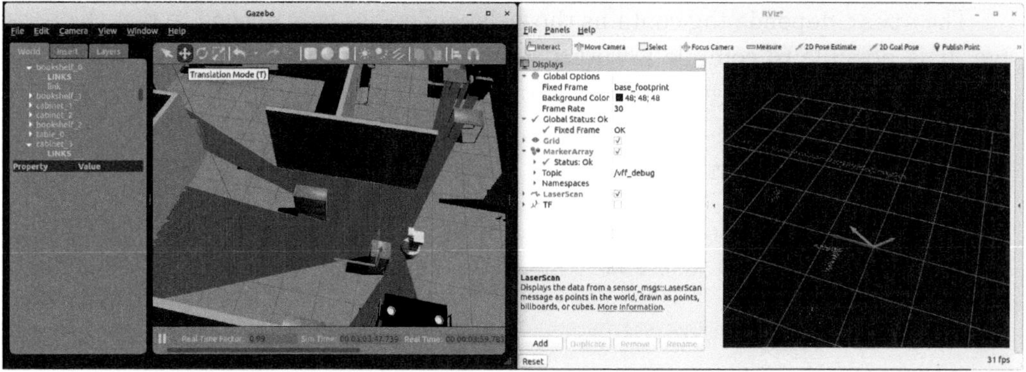

Figure 5.3: Execution of avoidance behavior.

5.1.7 Testing During Development

The code shown in the previous sections may contain calculation errors that can be detected before running it on a real robot and even before running it on a simulator. A very convenient strategy is, taking some (not all) concepts of test-driven development, doing tests simultaneously as the code is developed. This strategy has several advantages:

- Ensure that once a part of the software has been tested, other parts' changes do not negatively affect what has already been developed. The tests are incremental. All tests are always passed, assessing the new functionality and the validity of previously existing code, making development faster.
- The revision task is greatly simplified if the package receives contributions from other developers. Activating a CI (Continous Integration) system in your repository allows that each contribution has to compile correctly and pass all the tests, both functional and stylish. In this way, the reviewer focuses on verifying that the code does its job correctly.
- Many quality assurance procedures require the software to be tested. Saying "I will do the tests when I finish" is a fallacy: *You will not do them*, or it will be a tedious process that will not help you, so they are likely to be incomplete and ineffective.

ROS2 provides many testing tools that we can use easily. Let's start with the unit tests. ROS2 uses GoogleTest[4] to test C++ code. In order to use tests in the package, include some packages in the `package.xml`:

package.xml

```
<test_depend>ament_lint_auto</test_depend>
<test_depend>ament_lint_common</test_depend>
<test_depend>ament_cmake_gtest</test_depend>
```

[4]https://github.com/google/googletest

The `<test_depend>` tag contains those dependencies only needed to test the package. It is possible to compile a workspace, in this case only the package, excluding the tests, so these packages will not be taken into account in the dependencies:

```
$ colcon build --symlink-install --packages-select br2_vff_avoidance
--cmake-args -DBUILD_TESTING=off
```

As shown in the package structure, there is a `tests` directory with a C++ file (`vff_test.cpp`) that contains tests. To compile it, these sentences should be in `CMakeLists.txt`:

CMakeLists.txt

```
if(BUILD_TESTING)
  find_package(ament_lint_auto REQUIRED)
  ament_lint_auto_find_test_dependencies()

  set(ament_cmake_cpplint_FOUND TRUE)
  ament_lint_auto_find_test_dependencies()

  find_package(ament_cmake_gtest REQUIRED)
  add_subdirectory(tests)
endif()
```

tests/CMakeLists.txt

```
ament_add_gtest(vff_test vff_test.cpp)
ament_target_dependencies(vff_test ${dependencies})
target_link_libraries(vff_test ${PROJECT_NAME})
```

Once introduced the testing infrastructure in a package, see how to do unit tests. While developing the method `AvoidanceNode::get_vff` it is possible to check that it works correctly. Just create several synthetic `sensor_msgs::msg::LaserScan` messages and then check that this function returns correct values in all cases. In this file, it has been developed eight different cases. Let's see some of them:

tests/vff_test.cpp

```
sensor_msgs::msg::LaserScan get_scan_test_1(rclcpp::Time ts)
{
  sensor_msgs::msg::LaserScan ret;
  ret.header.stamp = ts;
  ret.angle_min = -M_PI;
  ret.angle_max = M_PI;
  ret.angle_increment = 2.0 * M_PI / 16.0;
  ret.ranges = std::vector<float>(16, std::numeric_limits<float>::infinity());

  return ret;
}

sensor_msgs::msg::LaserScan get_scan_test_5(rclcpp::Time ts)
{
  sensor_msgs::msg::LaserScan ret;
  ret.header.stamp = ts;
  ret.angle_min = -M_PI;
  ret.angle_max = M_PI;
  ret.angle_increment = 2.0 * M_PI / 16.0;
  ret.ranges = std::vector<float>(16, 5.0);
  ret.ranges[10] = 0.3;

  return ret;
}
```

Each function returns a sensor_msgs::msg::LaserScan message as if it had been generated by a laser with 16 different values, regularly distributed in the range $[-\pi, \pi]$. In get_scan_test_1 it simulates the case that no obstacles are detected in any case. At get_scan_test_5 it simulates that there is an obstacle at position 10, which corresponds to angle $-\pi + 10 * \frac{2\pi}{16} = 0.785$.

In order to access the method to be tested, since it is not public, it is convenient to make it protected and implement a class to access these functions:

tests/vff_test.cpp

```
class AvoidanceNodeTest : public br2_vff_avoidance::AvoidanceNode
{
public:
  br2_vff_avoidance::VFFVectors
  get_vff_test(const sensor_msgs::msg::LaserScan & scan)
  {
    return get_vff(scan);
  }

  visualization_msgs::msg::MarkerArray
  get_debug_vff_test(const br2_vff_avoidance::VFFVectors & vff_vectors)
  {
    return get_debug_vff(vff_vectors);
  }
};
```

It is possible to have all the needed tests in the same file. Each of them is defined using the macro TEST(id, sub_id), and inside, as if it were a function, write a program whose objective is to test the functionality of the code. In the case of get_vff, these are the **unitary tests**:

tests/vff_test.cpp

```
TEST(vff_tests, get_vff)
{
  auto node_avoidance = AvoidanceNodeTest();

  rclcpp::Time ts = node_avoidance.now();

  auto res1 = node_avoidance.get_vff_test(get_scan_test_1(ts));
  ASSERT_EQ(res1.attractive, std::vector<float>({1.0f, 0.0f}));
  ASSERT_EQ(res1.repulsive, std::vector<float>({0.0f, 0.0f}));
  ASSERT_EQ(res1.result, std::vector<float>({1.0f, 0.0f}));

  auto res2 = node_avoidance.get_vff_test(get_scan_test_2(ts));
  ASSERT_EQ(res2.attractive, std::vector<float>({1.0f, 0.0f}));
  ASSERT_NEAR(res2.repulsive[0], 1.0f, 0.00001f);
  ASSERT_NEAR(res2.repulsive[1], 0.0f, 0.00001f);
  ASSERT_NEAR(res2.result[0], 2.0f, 0.00001f);
  ASSERT_NEAR(res2.result[1], 0.0f, 0.00001f);

  auto res5 = node_avoidance.get_vff_test(get_scan_test_5(ts));
  ASSERT_EQ(res5.attractive, std::vector<float>({1.0f, 0.0f}));
  ASSERT_LT(res5.repulsive[0], 0.0f);
  ASSERT_LT(res5.repulsive[1], 0.0f);
  ASSERT_GT(atan2(res5.repulsive[1], res5.repulsive[0]), -M_PI);
  ASSERT_LT(atan2(res5.repulsive[1], res5.repulsive[0]), -M_PI_2);
  ASSERT_LT(atan2(res5.result[1], res5.result[0]), 0.0);
  ASSERT_GT(atan2(res5.result[1], res5.result[0]), -M_PI_2);
}

int main(int argc, char ** argv)
{
  rclcpp::init(argc, argv);

  testing::InitGoogleTest(&argc, argv);
  return RUN_ALL_TESTS();
}
```

The ASSERT_* macros check the expected values based on the input. ASSERT_EQ verifies that the two values are equal. When comparing floats, it is preferable to use ASSERT_NEAR, which checks that two values are equal with a specified margin in its third parameter. ASSERT_LT verifies that the first value is "Less Than" the second. ASSERT_GT verifies that the first value is "Greater Than" the second, and so on.

For example, case 5 (obstacle at angle 0.785) verifies that the coordinates of the repulsive vector are negative, both its angle is in the range $[-\pi, -\frac{\pi}{2}]$ (it is a vector opposite to angle 0.785) and that the resulting vector is in the range $[0, -\frac{\pi}{2}]$. If this is true, the algorithm is correct. Do these checks for each reading, with its expected values, and pay attention to extreme and unexpected cases, such as test 1.

It is also possible to do **integration tests**. Since nodes are objects, instantiate them and simulate their operation. For example, test the speeds published by AvoidanceNode when receiving the test messages. Let's see how to do it:

tests/vff_test.cpp

```
TEST(vff_tests, ouput_vels)
{
  auto node_avoidance = std::make_shared<AvoidanceNodeTest>();

  // Create a testing node with a scan publisher and a speed subscriber
  auto test_node = rclcpp::Node::make_shared("test_node");
  auto scan_pub = test_node->create_publisher<sensor_msgs::msg::LaserScan>(
    "input_scan", 100);

  geometry_msgs::msg::Twist last_vel;
  auto vel_sub = test_node->create_subscription<geometry_msgs::msg::Twist>(
    "output_vel", 1, [&last_vel] (geometry_msgs::msg::Twist::SharedPtr msg) {
      last_vel = *msg;
    });

  ASSERT_EQ(vel_sub->get_publisher_count(), 1);
  ASSERT_EQ(scan_pub->get_subscription_count(), 1);

  rclcpp::Rate rate(30);
  rclcpp::executors::SingleThreadedExecutor executor;
  executor.add_node(node_avoidance);
  executor.add_node(test_node);

  // Test for scan test #1
  auto start = node_avoidance->now();
  while (rclcpp::ok() && (node_avoidance->now() - start) < 1s) {
    scan_pub->publish(get_scan_test_1(node_avoidance->now()));
    executor.spin_some();
    rate.sleep();
  }
  ASSERT_NEAR(last_vel.linear.x, 0.3f, 0.0001f);
  ASSERT_NEAR(last_vel.angular.z, 0.0f, 0.0001f);

  // Test for scan test #2
}
```

1. Create an AvoidanceNodeTest (AvoidanceNode is also possible) node to test it.

2. Make a generic node called test_node to create a laser scan publisher and a speed subscriber.

3. When creating the speed subscriber, a lambda function has especified as a callback. This lambda function accesses the last_vel variable to update it with the last message received in the topic output_vel.

4. Create an executor and add both nodes to it to execute them.

5. During a second post at 30 Hz on input_scan a sensor reading corresponding to the synthetic readings.

6. In the end, verify that the published speeds are correct.

To run just these gtest tests, do it by running the binary that is in the tests directory of the package, in the build directory:

```
$ cd /bookros2_ws
$ build/br2_vff_avoidance/tests/vff_test
[==========] Running 2 tests from 1 test case.
[----------] Global test environment set-up.
[----------] 2 tests from vff_tests
[ RUN      ] vff_tests.get_vff
[       OK ] vff_tests.get_vff (18 ms)
[ RUN      ] vff_tests.ouput_vels
[       OK ] vff_tests.ouput_vels (10152 ms)
[----------] 2 tests from vff_tests (10170 ms total)

[----------] Global test environment tear-down
[==========] 2 tests from 1 test case ran. (10170 ms total)
[ PASSED   ] 2 tests.
```

To run all the tests for this package, even the style ones, use colcon:

```
$ colcon test --packages-select br2_vff_avoidance
```

If the test has finished with failures, go to check what has failed to the directory log/latest_test/br2_vff_avoidance/stdout_stderr.log. At the end of the file, there is a summary of the failed tests. For example, this message at the end indicates that tests 3, 4, 5, and 7 failed (errors were intentionally added for this explanation):

log/latest_test/br2_vff_avoidance/stdout_stderr.log

```
56% tests passed, tests failed out of 9

Label Time Summary:
copyright     =   0.37 sec*proc (1 test)
cppcheck      =   0.44 sec*proc (1 test)
cpplint       =   0.45 sec*proc (1 test)
flake8        =   0.53 sec*proc (1 test)
gtest         =  10.22 sec*proc (1 test)
lint_cmake    =   0.34 sec*proc (1 test)
linter        =   3.88 sec*proc (8 tests)
pep257        =   0.38 sec*proc (1 test)
uncrustify    =   0.38 sec*proc (1 test)
xmllint       =   0.99 sec*proc (1 test)

Total Test time (real) =  14.11 sec

The following tests FAILED:
        [  3 - cpplint (Failed)]
        [  4 - flake8 (Failed)]
        [  5 - lint_cmake (Failed)]
        [  7 - uncrustify (Failed)]
Errors while running CTest
```

Each line in this file begins with the section number corresponding to a test. Go, for example, to sections 3, 4, and 7 to see some of these errors:

log/latest_test/br2_vff_avoidance/stdout_stderr.log

```
3: br2_vff_avoidance/tests/vff_test.cpp:215:  Add #include <memory> for
   make_shared<>  [build/include_what_you_use] [4]
3: br2_vff_avoidance/include/br2_vff_avoidance/AvoidanceNode.hpp:15:  #ifndef header
   guard has wrong style, please use: BR2_VFF_AVOIDANCE__AVOIDANCENODE_HPP_
   [build/header_guard] [5]

4: ./launch/avoidance_vff.launch.py:34:3: E111 indentation is not a multiple of four
4:   ld.add_action(vff_avoidance_cmd)
4:   ^

7: --- src/br2_vff_avoidance/AvoidanceNode.cpp
7: +++ src/br2_vff_avoidance/AvoidanceNode.cpp.uncrustify
7: @@ -100,2 +100 @@
7: -   if (distance_min < OBSTACLE_DISTANCE)
7: -   {
7: +   if (distance_min < OBSTACLE_DISTANCE) {
7: @@ -109 +108 @@
7: -     vff_vector.repulsive[0] = cos(oposite_angle)*complementary_dist;
7: +     vff_vector.repulsive[0] = cos(oposite_angle) * complementary_dist;
7:
7: Code style divergence in file 'tests/vff_test.cpp':
```

- The errors in Section 3 correspond to cpplint, a C++ linter. The first error indicates that a header must be added since there are functions that are declared in it. In the second, it indicates that the style of the header guard in `AvoidanceNode.hpp` is incorrect, indicating which one should be used.
- The errors in section 4 correspond to flake8, a Python linter. This error indicates that the launcher file uses an incorrect indentation since it should be space, multiples of 4.
- The errors labeled with 7 correspond to uncrustify, another C++ linter. In a format similar to the output of the `diff` command, it tells the difference between the code that is written and the one that should be in good style. In this case, it indicates that the start of an `if` block on line 100 of `AvoidanceNode.cpp` should be on the same line as `if`. The second error indicates that there should be spaces on both sides of an operator.

The first time facing solving style problems, it can seem like a daunting task without much meaning. You would wonder why the style that it indicates is better than yours. Indeed you have been using this style for years, and you are very proud of how your source code looks like. You will not understand why you have to use two spaces in C++ to indent, and not the tab, for example, or why open the blocks in the same line of a `while` if you always opened in the next line.

The first reason is that it indicates a good style. Cpplint, for example, uses the Google C++ Style Guide[5], which is a widely accepted style guide adopted by most software development companies.

[5]https://google.github.io/styleguide/cppguide.html

The second is because you have to follow this style if you want to contribute to a ROS2 project or repository. Rarely a repository that accepts contributions does not have a continuous integration system that passes these tests. Imagine that you are the one who maintains a project. You'll want all of your code to have a consistent style. It would be a nightmare to make your own style guide or discuss with each contributor at every pull request style issues rather to focus on their contribution. The worst discussion I can recall with a colleague was using tabs against spaces. It is a discussion that will have no solution because it is like talking about religions. Using a standard solves these problems.

Furthermore, the last reason is that it will make you a better programmer. Most of the style rules have a practical reason. Over time, you will automatically apply the style you have corrected so many times when passing the tests, and your code will have a good style as you write it.

5.2 TRACKING OBJECTS

This section analyzes a project that contains other reactive behavior. In this case, the behavior tracks the objects that match a specific color with the robot's head.

There are several new concepts that are introduced in this project:

- **Image analysis**: So far, we have used a relatively simple sensor. Images provide more complex perceptual information from which a lot of information can be extracted. Remember that there is an essential part of Artificial Intelligence that deals with Artificial Vision, and it is one of the primary sensors in robots. We will show how to process these images with OpenCV, the reference library in this area.

- **Control at joint level**: In the previous projects, the commands were speeds sent to the robot. In this case, we will see how to command positions directly to the joints of the robot's neck.

- **Lifecycle Nodes**: ROS2 provides a particular type of Node called Lifecycle Node. This node is very useful to control the life cycle, including its startup, activation, and deactivation.

5.2.1 Perception and Actuation Models

This project uses the images from the robot's camera as a source of information. Whenever a node transmits an (non-compressed) image in ROS2 it uses the same type of message: `sensor_msgs/msg/Image`. All the drivers of all cameras supported in ROS2 use it. See what the message format is:

```
$ ros2 interface show sensor_msgs/msg/image

# This message contains an uncompressed image
# This message contains an uncompressed

std_msgs/Header header  # Header timestamp should be acquisition time of image
                        # Header frame_id should be optical frame of camera
                        # origin of frame should be optical center of camera
                        # +x should point to the right in the image
                        # +y should point down in the image
                        # +z should point into to plane of the image
                        # If the frame_id and the frame_id of the CameraInfo
                        # message associated with the image conflict
                        # the behavior is undefined

uint32 height           # image height, that is, number of rows
uint32 width            # image width, that is, number of columns

# The legal values for encoding are in file src/image_encodings.cpp
# If you want to standardize a new string format, join
# ros-users@lists.ros.org and send an email proposing a new encoding.

string encoding   # Encoding of pixels -- channel meaning, ordering, size
                  # from the list in include/sensor_msgs/image_encodings.hpp

uint8 is_bigendian # is this data bigendian?
uint32 step        # Full row length in bytes
uint8[] data       # actual matrix data, size is (step * rows)
```

Camera drivers often publish (only once, in transient local QoS) information about camera parameters as a `sensor_msgs/msg/CameraInfo` message, which includes intrinsic and distortion parameters, projection matrix, and more. With this information, we can work with stereo images, for example, or we can combine this information with a depth image to reconstruct the 3D scene. The process of calibrating a camera[6] has to do with calculating the values that are published in this message. A good exercise is reading this message format, although it is not used in this chapter.

Although it is possible to use a simple `sensor_msgs/msg/Image` publisher or subscriber, it is usual when working with images using different transport strategies (compression, streaming codecs ...) using specific publishers/subscribers. The developer uses them and ignores how the images are transported – he just sees an `sensor_msgs/msg/Image`. Check available transport plugins typing:

[6]http://wiki.ros.org/image_pipeline

```
$ ros2 run image_transport list_transports

Declared transports:
image_transport/compressed
image_transport/compressedDepth
image_transport/raw
image_transport/theora

Details:
----------
...
```

Run the simulated Tiago and check the topics to see that there is more than one topic for 2D images:

```
$ ros2 topic list

/head_front_camera/image_raw/compressed
/head_front_camera/image_raw/compressedDepth
/head_front_camera/image_raw/theora
/head_front_camera/rgb/camera_info
/head_front_camera/rgb/image_raw
```

The developer has not created all these topics one by one, but has used an `image_transport::Publisher` that has generated all these topics taking into account the available transport plugins. In the same way, to obtain the images, it is convinient to use a `image_transport::Subscriber`, as we will see below. Using compressed images may be good if the image is big or the network reliability is not the best. The trade-off is a bit more CPU load on the source and destination.

The image message format is for transporting images, not for processing them. It is not common to work directly with images as raw byte sequences. The usual way is to use some image processing library, and the most widely used is OpenCV[7]. OpenCV provides several hundreds of computer vision algorithms.

The main data type that OpenCV uses to work with images is `cv::Mat`. ROS2 provides tools to transform `sensor_msgs/msg/Image` into `cv::Mat`, and vice versa:

```
void image_callback(const sensor_msgs::msg::Image::ConstSharedPtr & msg)
{
  cv_bridge::CvImagePtr cv_ptr;
  cv_ptr = cv_bridge::toCvCopy(msg, sensor_msgs::image_encodings::BGR8);
  cv::Mat & image_src = cv_ptr->image;

  sensor_msgs::msg::Image image_out = *cv_ptr->toImageMsg();
}
```

In the perception model of our project, the segmentation of an image will be done by color. It is convenient to work in HSV[8], instead of RGB, which is the encoding in which we receive the messages. HSV encoding represents a pixel in color with three components: Hue, Saturation, and Value. Working in HSV allows us to establish color ranges more robustly to lighting changes since this is what the V component is mainly

[7]https://docs.opencv.org/5.x/d1/dfb/intro.html
[8]https://en.wikipedia.org/wiki/HSL_and_HSV

responsible for, and if the range is wider, we can continue to detect the same color even if the illumination changes.

The following code transforms a `cv::mat` to HSV and calculates an image mask with the pixels that match the color of the furniture in the default simulated world of Tiago in Gazebo, as shown in Figure 5.4:

```
cv::Mat img_hsv;
cv::cvtColor(cv_ptr->image, img_hsv, cv::COLOR_BGR2HSV);

cv::Mat1b filtered;
cv::inRange(img_hsv, cv::Scalar(15, 50, 20), cv::Scalar(20, 200, 200), filtered);
```

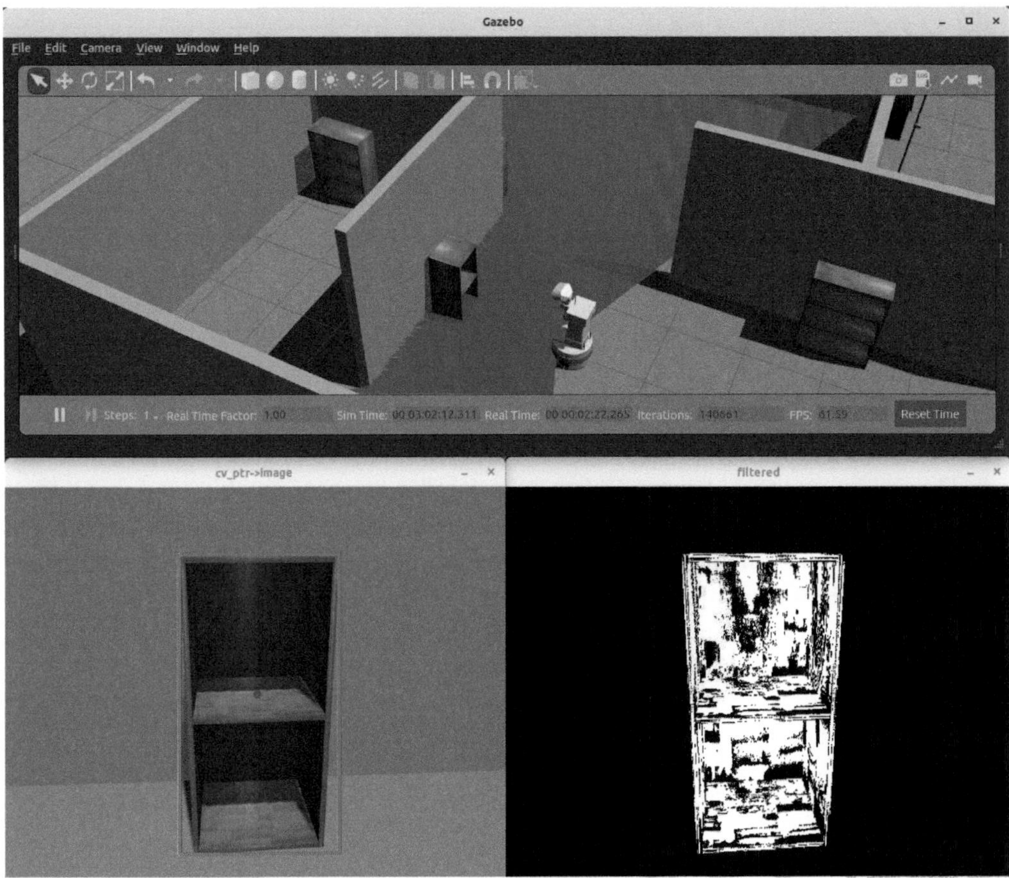

Figure 5.4: Object detection by color using an HSV range filter.

Finally, the output of the processing of an image in this project is a message of type `vision_msgs/msg/Detection2D` (examine the fields in this message for yourself), from which we use its `header`, `bbox`, and `source_img field`. It is not required to use all the fields. The original image is included to have the image's dimensions where the detection is made, whose importance will be shown below.

The **action model** is a position control of the robot head. The robot has two joints that control the position the camera is pointing at: `head_1_joint` for horizontal control (pan) and `head_2_joint` for vertical control (tilt).

In ROS2 the control of the joints is done by the framework **ros2_control**[9]. The developers of the simulated Tiago robot have used a trajectory controller (joint_trajectory_controller) for the two joints of the robot's neck. Through two topics (as shown in Figure 5.5), it allows reading the state of the joints, and sending commands in the form of a set of waypoints (Figure 5.6) to be reached at specific time instants. Waypoints consist of positions and optionally velocities, accelerations, and effort, as well as a time from start to be applied.

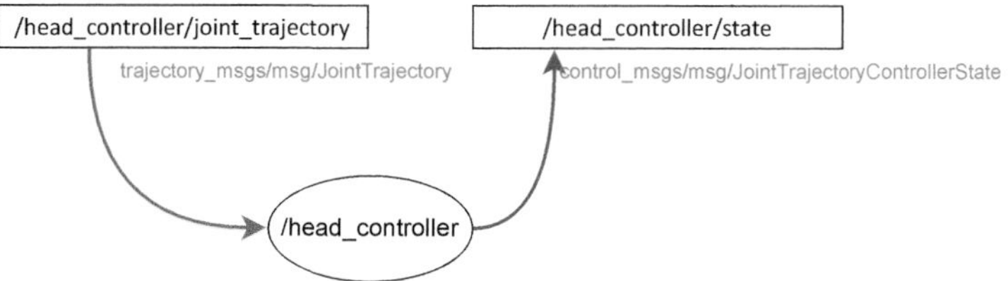

Figure 5.5: Head controller topics.

trajectory_msgs/msg/JointTrajectory

joint_names[]	Header	points[]														
	head_1_joint	pos	vel	acc	effo	time_from start	pos	vel	acc	effo	time_from start	pos	vel	acc	effo	time_from start
	head_2_joint	pos	vel	acc	effo	time_from start	pos	vel	acc	effo	time_from start	pos	vel	acc	effo	time_from start
		WP 0					WP 1					WP 2				

Figure 5.6: `trajectory_msgs/msg/JointTrajectory` message format.

Obtaining the 3D position of the object to the robot and calculating the position of the neck joints to center it in the image would probably be an adequate solution in a real active vision system, but quite complex at the moment. We will simply implement a control in the image domain.

The node that controls the robot's neck receives two values (called error) that indicate the difference between the current position and the desired position for pan and tilt, respectively. If a value is 0, it indicates that it is in the desired position. If it is less than 0, the joint has to move in one direction, and greater than zero has to move in the other direction. The range of values is $[-1, +1]$ for each joint, as shown in Figure 5.7. As this node performs iterative control and neck movements can be very fast, a PID controller will control the position to which each joint is commanded to correct its speed.

[9]http://control.ros.org/index.html

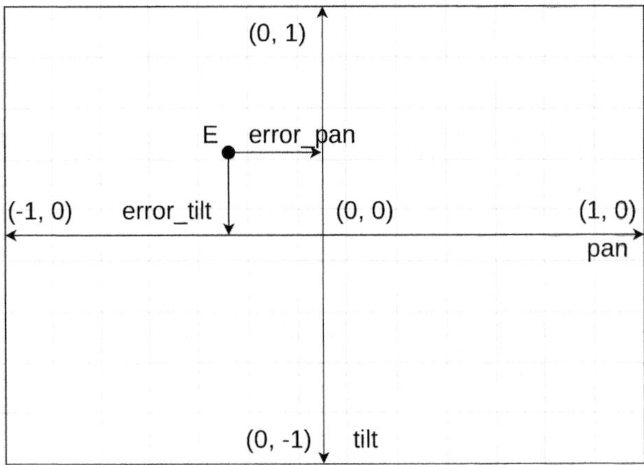

Figure 5.7: Diagram for pan/tilt control. `E` indicates the desired position. `error_*` indicates the difference between the current position and the desired pan/tilt position.

5.2.2 Computation Graph

The Computation Graph of this project (Figure 5.8) shows how this problem is divided into three nodes within the same process. The reason is that each node (`ObjectDetector` and `HeadController`) can be executed separately, and be reused in other problems (we will do it in next chapters). Each one has been designed in this way to be reusable, with inputs and outputs that try to be generic, not strongly coupled to this problem.

In this Computation Graph, the `HeadController` has been represented differently from the rest of the nodes. This node will be implemented as **LifeCycle Node**, which we will explain in the Section 5.2.3. For now, we will say that it is like a standard node but that it can be activated and deactivated during its operation.

The `HeadController` receives a pan/tilt speed, each in the range $[-1, 1]$. Note that since there is no standard ROS2 message that fits our problem (we could have used `geometry_msgs/msg/Pose2D`, ignoring the field theta), we have created a custom `br2_tracking_msgs/msg/PanTiltCommand` message containing the needed information. We will see below how we have done to create our custom message.

The `ObjectDetector` publishes, for each image, the result of the detection of the furniture in the image. It will return the coordinate, in pixels, of the detection, as well as the bounding box of the object.

The output of the `ObjectDetector` does not completely match the input of the `HeadController`. `ObjectDetector` publishes its output in pixels. In this case, the image resolution is 640×480 so its range is $[0, 640]$ for the horizontal X component and $[0, 480]$ for the vertical Y component. Therefore, we create a node, `tracker`, with a straightforward task, which is to adapt the output of the `ObjectDetector` to the input of the `HeadController`, to make a control in the image, moving the head so that the detected object is always in the center of the image.

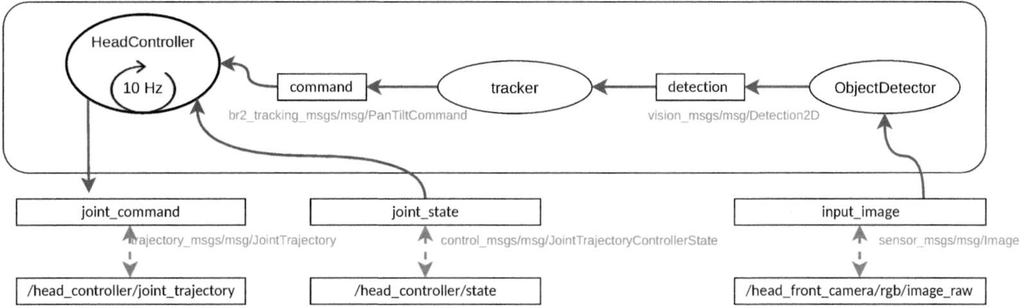

Figure 5.8: Computation Graph for Object Tracking project.

5.2.3 Lifecycle Nodes

So far, we have seen that the nodes in ROS2 are objects of class `Node` that inherit methods that allow us to communicate with other nodes or obtain information. In ROS2, there is a type of node, the **LifeCycleNode**, whose lifetime is defined using states and the transitions between them:

- When a LifeCycleNode is created, it is in *Unconfigured* state, and it must trigger the *configure* transition to enter the *Inactive* state.
- A LifeCycleNode is working when it is in the *Active* state, from which it can transition from the *Inactive* state through the *activate* transition. It is also possible to transition from the *Active* to *Inactive* state through the *deactivate* transition.
- The necessary tasks and checks can be performed at each transition. Even a transition can fail and not transit if the conditions specified in the code of its transition are not met.
- In case of error, the node can go to *Finalized* state.
- When a node has completed its task, it can transition to *Finalized.*

See a diagram of these states and transitions in Figure 5.9.
Lifecycle nodes provide a node execution model that allows:

- Make them predictable. For example, in ROS2, the parameters should be read only in the *configuring* transition.
- When there are multiple Nodes, we can coordinate their startup. We can define that specific nodes are not activated until they are configured. We also can specify some orders in the startup.
- Programmatically, it allows having another option beyond the constructor to start its components. Remember that in C++, a Node is not completely built until its constructor has finished. This usually brings problems if we require a `shared_ptr` to `this`.

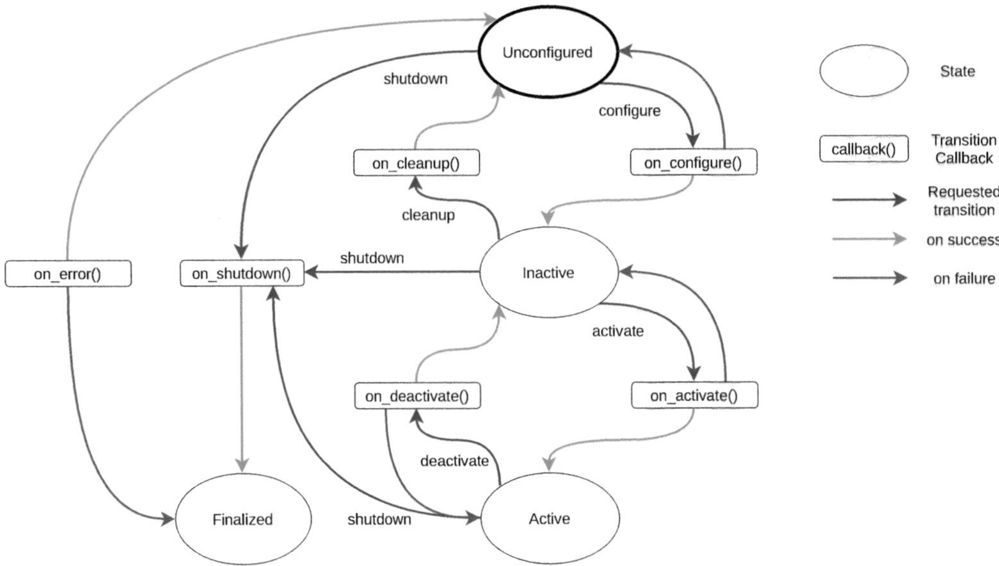

Figure 5.9: Diagram of states and transitions in Lifecycle Nodes.

An example could be a sensor driver. If the physical device cannot be accessed, it cannot transit to the *Inactive* state. In addition, all the initial setup time of the device would be set to this state so that its activation would be immediate. Another example is the startup of a robot driver. It would not boot until all its sensor/actuator nodes are in the *Active* state.

5.2.4 Creating Custom Messages

We have previously specified that the input of node `HeadController` is of type `br2_tracking_msgs/msg/PanTiltCommand` because there was no type of message that conformed to what we needed. One golden rule in ROS2 is *not to create a message if there is already a standard available*, as we can benefit from available tools for this message. In this case, no standard will serve our purposes. In addition, it is the perfect excuse to show how to create custom messages.

First of all, *when creating new messages* (new interfaces, in general), even in the context of a specific package, it is highly recommended that you *make a separate package*, ending in `_msgs`. Tools may exist in the future that needs to receive messages of this new type, but we do not necessarily have to depend on the packages for which they were created.

Next we show the structure of package `br2_tracking_msgs` that contains only the definition of message `br2_tracking_msgs/msg/PanTiltCommand`:

Package br2_tracking_msgs

```
br2_tracking_msgs/
├── CMakeLists.txt
├── msg
│   └── PanTiltCommand.msg
└── package.xml
```

Packages that contain interfaces have, besides a package.xml and a CMakeLists.txt, a directory for each type of interface (message, service, or action) that is being defined. In our case, it is a message, so we will have a msg directory that contains a .msg file for each new message to define. Let's see the the definition of the PanTiltCommand message:

msg/PanTiltCommand.msg

```
float64 pan
float64 tilt
```

The important part in the CMakeLists.txt is the rosidl_generate_interfaces statement, in which we specify where the interface definitions are:

CMakeLists.txt

```
find_package(ament_cmake REQUIRED)
find_package(builtin_interfaces REQUIRED)
find_package(rosidl_default_generators REQUIRED)

rosidl_generate_interfaces(${PROJECT_NAME}
  "msg/PanTiltCommand.msg"
  DEPENDENCIES builtin_interfaces
)

ament_export_dependencies(rosidl_default_runtime)
ament_package()
```

5.2.5 Tracking Implementation

The structure of the br2_tracking package, shown below, follows the guidelines already recommended above.

Package br2_vff_avoidance

```
br2_tracking
├── CMakeLists.txt
├── config
│   └── detector.yaml
├── include
│   └── br2_tracking
│       ├── HeadController.hpp
│       ├── ObjectDetector.hpp
│       └── PIDController.hpp
├── launch
│   └── tracking.launch.py
├── package.xml
├── src
│   ├── br2_tracking
│   │   ├── HeadController.cpp
│   │   ├── ObjectDetector.cpp
│   │   └── PIDController.cpp
│   └── object_tracker_main.cpp
└── tests
    ├── CMakeLists.txt
    └── pid_test.cpp
```

- The HeadController and ObjectDetector nodes will be compiled as libraries independently of the main program object_tracker_main.cpp. The latter will be in src, while the nodes will have their headers in include/br2_tracking and their definitions in src/br2_tracking.

- The library also includes a class to use PID controllers, used in HeadController.
- A launcher will launch the executable with the necessary parameters and remaps.
- There is a config directory that contains a YAML file with the HSV range that ObjectDetector will use to detect in the image the furniture of Tiago's default stage in Gazebo.
- The tests directory includes tests for the PID controller.

The reader will have noticed that there is no file for a tracker node in this structure. This node, being so simple, has been implemented in object_tracker_main.cpp as follows:

src/object_tracker_main.cpp

```
auto node_detector = std::make_shared<br2_tracking::ObjectDetector>();
auto node_head_controller = std::make_shared<br2_tracking::HeadController>();
auto node_tracker = rclcpp::Node::make_shared("tracker");

auto command_pub = node_tracker->create_publisher<br2_tracking_msgs::msg::PanTiltCommand>(
  "/command", 100);
auto detection_sub = node_tracker->create_subscription<vision_msgs::msg::Detection2D>(
  "/detection", rclcpp::SensorDataQoS(),
  [command_pub](vision_msgs::msg::Detection2D::SharedPtr msg) {
    br2_tracking_msgs::msg::PanTiltCommand command;
    command.pan = (msg->bbox.center.x / msg->source_img.width) * 2.0 - 1.0;
    command.tilt = (msg->bbox.center.y / msg->source_img.height) * 2.0 - 1.0;
    command_pub->publish(command);
  });

rclcpp::executors::SingleThreadedExecutor executor;
executor.add_node(node_detector);
executor.add_node(node_head_controller->get_node_base_interface());
executor.add_node(node_tracker);
```

node_tracker is a generic ROS2 node, from which we construct a publisher to the /command topic, and a subscriber to the /detection. We have specified the subscriber callback as a lambda function that takes from the input message the position in pixels of the detected object, together with the size of the image, and generates the inputs for node HeadController, following the scheme already shown in Figure 5.7.

Notice that when adding the node node_head_controller to executor, we have used the get_node_base_interface method. This is because it is a LifeCycleNode, as we introduced earlier, and add_node does not yet support adding this type of node directly. Fortunately, we can do it through a basic interface supported by LifeCycleNode and regular nodes using this method.

The ObjectDetector will be a rclcpp::Node, with an image subscriber (using image_transport) and a 2D detection message publisher. There are two member variables that will be used in the detection process.

include/br2_tracking/ObjectDetector.hpp

```
class ObjectDetector : public rclcpp::Node
{
public:
  ObjectDetector();

  void image_callback(const sensor_msgs::msg::Image::ConstSharedPtr & msg);

private:
  image_transport::Subscriber image_sub_;
  rclcpp::Publisher<vision_msgs::msg::Detection2D>::SharedPtr detection_pub_;

  // HSV ranges for detection [h - H] [s - S] [v - V]
  std::vector<double> hsv_filter_ranges_ {0, 180, 0, 255, 0, 255};
  bool debug_ {true};
};
```

These variables, with a default value, will be initialized using parameters. They are the HSV color ranges and a variable that, by default, causes a window to be displayed with the detection result for debugging purposes.

src/br2_tracking/ObjectDetector.cpp

```
ObjectDetector::ObjectDetector()
: Node("object_detector")
{
  declare_parameter("hsv_ranges", hsv_filter_ranges_);
  declare_parameter("debug", debug_);

  get_parameter("hsv_ranges", hsv_filter_ranges_);
  get_parameter("debug", debug_);
}
```

When executing the program with all the nodes, a parameter file in the config directory will be specified to set the color filter.

config/detector.yaml

```
/object_detector:
  ros__parameters:
    debug: true
    hsv_ranges:
      - 15.0
      - 20.0
      - 50.0
      - 200.0
      - 20.0
      - 200.0
```

This node is designed to obtain a result for each image that arrives, so the processing is done directly in the callback, as long as there is a subscriber to this result.

Creating an `image_transport::Subscriber` is very similar to a `rclcpp::Subscription`. The first parameter is a `rclcpp::Node*`, so we use `this`. The fourth parameter indicates the transport method, in this case raw. We adjust the quality of service in the last parameters to the usual in sensors.

src/br2_tracking/ObjectDetector.cpp

```
ObjectDetector::ObjectDetector()
: Node("object_detector")
{
  image_sub_ = image_transport::create_subscription(
    this, "input_image", std::bind(&ObjectDetector::image_callback, this, _1),
    "raw", rclcpp::SensorDataQoS().get_rmw_qos_profile());

  detection_pub_ = create_publisher<vision_msgs::msg::Detection2D>("detection", 100);
}

void
ObjectDetector::image_callback(const sensor_msgs::msg::Image::ConstSharedPtr & msg)
{
  if (detection_pub_->get_subscription_count() == 0) {return;}
  ...

  vision_msgs::msg::Detection2D detection_msg;
  ...
  detection_pub_->publish(detection_msg);
}
```

Image processing was already introduced in the previous sections. Once the image message has been transformed to a `cv::Mat`, we proceed to transform it from RGB to HSV, and we do a color filter. The `cv::boundingRect` function calculates a bounding box from the mask resulting from the color filtering. The `cv::moments` function calculates the center of mass of these pixels.

src/br2_tracking/ObjectDetector.cpp

```
const float & h = hsv_filter_ranges_[0];
const float & H = hsv_filter_ranges_[1];
const float & s = hsv_filter_ranges_[2];
const float & S = hsv_filter_ranges_[3];
const float & v = hsv_filter_ranges_[4];
const float & V = hsv_filter_ranges_[5];

cv_bridge::CvImagePtr cv_ptr;
try {
  cv_ptr = cv_bridge::toCvCopy(msg, sensor_msgs::image_encodings::BGR8);
} catch (cv_bridge::Exception & e) {
  RCLCPP_ERROR(get_logger(), "cv_bridge exception: %s", e.what());
  return;
}

cv::Mat img_hsv;
cv::cvtColor(cv_ptr->image, img_hsv, cv::COLOR_BGR2HSV);

cv::Mat1b filtered;
cv::inRange(img_hsv, cv::Scalar(h, s, v), cv::Scalar(H, S, V), filtered);

auto moment = cv::moments(filtered, true);
cv::Rect bbx = cv::boundingRect(filtered);

auto m = cv::moments(filtered, true);
if (m.m00 < 0.000001) {return;}
int cx = m.m10 / m.m00;
int cy = m.m01 / m.m00;

vision_msgs::msg::Detection2D detection_msg;
detection_msg.header = msg->header;
detection_msg.bbox.size_x = bbx.width;
detection_msg.bbox.size_y = bbx.height;
detection_msg.bbox.center.x = cx;
detection_msg.bbox.center.y = cy;
detection_msg.source_img = *cv_ptr->toImageMsg();
detection_pub_->publish(detection_msg);
```

In the previous code, the image is processed, the bounding box bbx of the filtered pixels in filtered is obtained, and it is published, together with the center of mass (cx, cy). In addition, the optional source_img field is filled in, since we require the size of the image in object_tracker_main.cpp.

The HeadController implementation is a bit more complex. Let's focus first on the fact that it is a Lifecycle node, and that its control loop is only called when it is active. Let's look at the declaration of the node, just the part of its control infrastructure:

include/br2_tracking/HeadController.hpp

```
class HeadController : public rclcpp_lifecycle::LifecycleNode
{
public:
  HeadController();

  CallbackReturn on_configure(const rclcpp_lifecycle::State & previous_state);
  CallbackReturn on_activate(const rclcpp_lifecycle::State & previous_state);
  CallbackReturn on_deactivate(const rclcpp_lifecycle::State & previous_state);

  void control_sycle();

private:
  rclcpp_lifecycle::LifecyclePublisher<trajectory_msgs::msg::JointTrajectory>::SharedPtr
    joint_pub_;
  rclcpp::TimerBase::SharedPtr timer_;
};
```

The LifecycleNode::create_subscription method returns an rclcpp_lifecycle:: LifecyclePublisher instead of an rclcpp::Publisher. Although its functionality is similar, it is necessary to activate it so that it can be used.

A LifeCycleNode can redefine the functions that are called when a transition between states is triggered in the derived class. These functions can return SUCCESS or FAILURE. If it returns SUCCESS, the transition is allowed. If FAILURE is returned, it is not transitioned to the new state. All of these methods return SUCCESS in the base class, but the developer can redefine them to establish the rejection conditions.

In this case, the transitions leading to the inactive state (on_configure) and those that transition between active and inactive (on_activate and on_deactivate) are redefined:

src/br2_tracking/HeadController.cpp

```
HeadController::HeadController()
: LifecycleNode("head_tracker")
{
  joint_pub_ = create_publisher<trajectory_msgs::msg::JointTrajectory>(
    "joint_command", 100);
}

CallbackReturn
HeadController::on_configure(const rclcpp_lifecycle::State & previous_state)
{
  return CallbackReturn::SUCCESS;
}
CallbackReturn
HeadController::on_activate(const rclcpp_lifecycle::State & previous_state)
{
  joint_pub_->on_activate();
  timer_ = create_wall_timer(100ms, std::bind(&HeadController::control_sycle, this));
```

src/br2_tracking/HeadController.cpp

```
  return CallbackReturn::SUCCESS;
}

CallbackReturn
HeadController::on_deactivate(const rclcpp_lifecycle::State & previous_state)
{
  joint_pub_->on_deactivate();
  timer_ = nullptr;

  return CallbackReturn::SUCCESS;
}

void
HeadController::control_sycle()
{
}
```

All previous transitions return SUCCESS, so all transitions are carried out. In the case of developing a laser driver, for example, some transition (configure or activate) would fail if the device is not found or cannot be accessed.

The above code has two aspects that are interesting to explain:

- The `control_cycle` method contains our control logic and is set to run at 10 Hz. Note that the timer is created at `on_activate`, which is when the active state is transitioned. Likewise, disabling this timer is simply destroying it by going inactive. This way `control_cycle` will not be called and the control logic will only be executed when the node is active.

- The publisher must be activated in `on_activate` and deactivated in `on_deactivate`.

The `HeadController` node will execute iteratively, receiving the current state of the neck joints through the topic `/joint_state`, and of the move commands through the `/command` topic. As usual in this schematic, both values in `last_state_` and `last_command_` are stored to be used when we execute the next cycle of the control logic. Also, the timestamp of the last received command is saved. When stopping receiving commands, the robot should return to the initial position.

include/br2_tracking/HeadController.hpp

```
class HeadController : public rclcpp_lifecycle::LifecycleNode
{
public:
  void joint_state_callback(
    control_msgs::msg::JointTrajectoryControllerState::UniquePtr msg);
  void command_callback(br2_tracking_msgs::msg::PanTiltCommand::UniquePtr msg);

private:
  rclcpp::Subscription<br2_tracking_msgs::msg::PanTiltCommand>::SharedPtr command_sub_;
  rclcpp::Subscription<control_msgs::msg::JointTrajectoryControllerState>::SharedPtr
    joint_sub_;
  rclcpp_lifecycle::LifecyclePublisher<trajectory_msgs::msg::JointTrajectory>::SharedPtr
    joint_pub_;

  control_msgs::msg::JointTrajectoryControllerState::UniquePtr last_state_;
  br2_tracking_msgs::msg::PanTiltCommand::UniquePtr last_command_;
  rclcpp::Time last_command_ts_;
};
```

src/br2_tracking/HeadController.cpp

```
void
HeadController::joint_state_callback(
  control_msgs::msg::JointTrajectoryControllerState::UniquePtr msg)
{
  last_state_ = std::move(msg);
}

void
HeadController::command_callback(br2_tracking_msgs::msg::PanTiltCommand::UniquePtr msg)
{
  last_command_ = std::move(msg);
  last_command_ts_ = now();
}
```

The format of `control_msgs::msg::JointTrajectoryControllerState` is designed to report the name of the controlled joints, as well as the desired, current, and error trajectories:

```
$ ros2 interface show control_msgs/msg/JointTrajectoryControllerState

std_msgs/Header header
string[] joint_names
trajectory_msgs/JointTrajectoryPoint desired
trajectory_msgs/JointTrajectoryPoint actual
trajectory_msgs/JointTrajectoryPoint error # Redundant, but useful
```

Using a `trajectory_msgs::msg::JointTrajectory` may seem complicated at first, but it is not if we analyze the following code, which is a command to put the robot's neck in the initial state while looking at Figure 5.6:

src/br2_tracking/HeadController.cpp

```
CallbackReturn
HeadController::on_deactivate(const rclcpp_lifecycle::State & previous_state)
{
  trajectory_msgs::msg::JointTrajectory command_msg;
  command_msg.header.stamp = now();
  command_msg.joint_names = last_state_->joint_names;
  command_msg.points.resize(1);
  command_msg.points[0].positions.resize(2);
  command_msg.points[0].velocities.resize(2);
  command_msg.points[0].accelerations.resize(2);
  command_msg.points[0].positions[0] = 0.0;
  command_msg.points[0].positions[1] = 0.0;
  command_msg.points[0].velocities[0] = 0.1;
  command_msg.points[0].velocities[1] = 0.1;
  command_msg.points[0].accelerations[0] = 0.1;
  command_msg.points[0].accelerations[1] = 0.1;
  command_msg.points[0].time_from_start = rclcpp::Duration(1s);

  joint_pub_->publish(command_msg);

  return CallbackReturn::SUCCESS;
}
```

- The `joint_names` field is a `std::vector<std::string>` containing the name of the joints being controlled. In this case, there are two, and they are the same ones that are already in the state message.
- A single waypoint will be sent (for this reason, the points field is resized to 1), in which a position, speed, and acceleration must be specified for each joint

(since there are two joints, each of these fields is resized to two). Position 0 corresponds to the joint that in `joint_names` is at 0, and so on.

- `time_from_start` indicates the time required to reach the commanded position. As it is the last command sent before deactivating (that is why its desired positions are 0), one second will be enough not to force the neck motors.

The controller of the neck joints is controlled by sending commands containing positions, but what is received from the `ObjectDetector` is the speed control that should be done to center the detected object in the image.

The first implementation could be to send as position, the current position combined with the received control:

src/br2_tracking/HeadController.cpp

```
command_msg.points[0].positions[0] = last_state_->actual.positions[0] - last_command_->pan;
command_msg.points[0].positions[1] = last_state_->actual.positions[1] -last_command_->tilt;
```

If the reader uses this implementation, he would see that if we want to be reactive enough, even if the difference between the `ObjectDetector` and `HeadDetector` frequencies were small, the robot's head might start to oscillate, trying to center the image on the detected object. It is difficult for the robot to maintain a stable focus on the detected object. This problem is solved in engineering using a PID controller, one per joint that limits the speed while also absorbing small unwanted oscillations of the neck.

include/br2_tracking/HeadController.hpp

```
class HeadController : public rclcpp_lifecycle::LifecycleNode
{
private:
  PIDController pan_pid_, tilt_pid_;
};
```

For each PID, define a value for the proportional component K_p, the integrating component K_i, and the derivative component K_d. Without going into details, since it is not the objective of this book to describe in-depth the underlying control theory, intuitively, the proportional component brings us closer to the objective. The integrator component compensates for persistent deviations that move us away from the objective. The derivative component tries to damp minor variations when close to the control objective.

Figure 5.10 shows a diagram of this PID controller.

The control command coming from the tracker is the value that the PID should try to keep at 0, so it is the error in t, $e(t)$. Each component of the PID is computed separately and then added to obtain the control to apply $u(t)$. The position sent to the joints will be the current position of the joint plus $u(t)$. The system feeds back since at $t+1$ the effect of the control is reflected in a change in the object's position in the image towards its center.

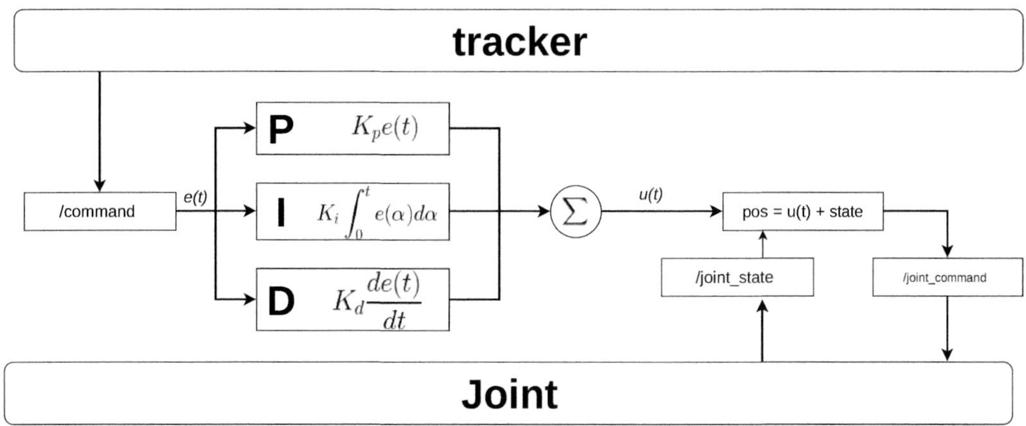

Figure 5.10: Diagram for PID for one joint.

Our PID starts by specifying four values: the minimum and maximum input reference expected in the PID and the minimum and maximum output produced. Negative input produces negative outputs:

src/br2_tracking/PIDController.hpp

```
class PIDController
{
public:
  PIDController(double min_ref, double max_ref, double min_output, double max_output);

  void set_pid(double n_KP, double n_KI, double n_KD);
  double get_output(double new_reference);
};
```

src/br2_tracking/HeadController.cpp

```
HeadController::HeadController()
: LifecycleNode("head_tracker"),
  pan_pid_(0.0, 1.0, 0.0, 0.3),
  tilt_pid_(0.0, 1.0, 0.0, 0.3)
{
}
CallbackReturn
HeadController::on_configure(const rclcpp_lifecycle::State & previous_state)
{
  pan_pid_.set_pid(0.4, 0.05, 0.55);
  tilt_pid_.set_pid(0.4, 0.05, 0.55);
}
void
HeadController::control_sycle()
{
  double control_pan = pan_pid_.get_output(last_command_->pan);
  double control_tilt = tilt_pid_.get_output(last_command_->tilt);

  command_msg.points[0].positions[0] = last_state_->actual.positions[0] - control_pan;
  command_msg.points[0].positions[1] = last_state_->actual.positions[1] - control_tilt;
}
```

5.2.6 Executing the Tracker

In the main program `object_tracker_main.cpp` all the nodes are created and added to an executor. Just before starting spinning the nodes, we trigger the configure transition for node node_head_controller. The node will be ready to be activated when requested.

src/object_tracker_main.cpp

```
rclcpp::executors::SingleThreadedExecutor executor;
executor.add_node(node_detector);
executor.add_node(node_head_controller->get_node_base_interface());
executor.add_node(node_tracker);

node_head_controller->trigger_transition(
  lifecycle_msgs::msg::Transition::TRANSITION_CONFIGURE);
```

A launcher remaps the topics and loads the file with the HSV filter parameters:

launch/tracking.launch.py

```
params_file = os.path.join(
  get_package_share_directory('br2_tracking'),
  'config',
  'detector.yaml'
  )

object_tracker_cmd = Node(
  package='br2_tracking',
  executable='object_tracker',
  parameters=[{
    'use_sim_time': True
  }, params_file],
  remappings=[
    ('input_image', '/head_front_camera/rgb/image_raw'),
    ('joint_state', '/head_controller/state'),
    ('joint_command', '/head_controller/joint_trajectory')
  ],
  output='screen'
)
```

Start the Tiago simulated (the home world, by default) gazebo:

```
$ ros2 launch br2_tiago sim.launch.py
```

In another terminal, launch the project:

```
$ ros2 launch br2_tracking tracking.launch.py
```

The detection windows are not shown until the first object is detected, but HeadController is in *Inactive* state, and no tracking will be done.

See how we can manage the LifeCycleNode at runtime, such as head_tracker (the name of the HeadController node). Keep our project running, with the robot tracking an object.

Using the following command, check what LifeCycle nodes are currently running:

```
$ ros2 lifecycle nodes

/head_tracker
```

Now verify the state it is currently in:

```
$ ros2 lifecycle get /head_tracker

inactive [3]
```

Good. The LifeCycleNode is in the *Inactive* state, just as expected. Obtain what transitions can be triggered from the current state:

```
$ ros2 lifecycle list /head_tracker

- cleanup [2]
            Start: inactive
            Goal: cleaningup
- activate [3]
            Start: inactive
            Goal: activating
- shutdown [6]
            Start: inactive
            Goal: shuttingdown
```

Activate the node to start tracking the detected object:

```
$ ros2 lifecycle set /head_tracker activate

Transitioning successful
```

Run a teleoperator in a third terminal to teleoperate the robot toward the furniture. Then, the robot will move (when `HeadController` is *Active*) the head to center the furniture in the image. As soon as the robot does not perceive the objects, it will move the head to the initial position:

```
$ ros2 run teleop_twist_keyboard teleop_twist_keyboard --ros-args -r
cmd_vel:=key_vel
```

Deactivate the node and check how the neck of the robot returns to its initial position. Remember that it was what was commanded in this transition, in the `on_deactivate` method.

```
$ ros2 lifecycle set /head_tracker deactivate

Transitioning successful
```

To activate it again, type:

```
$ ros2 lifecycle set /head_tracker activate

Transitioning successful
```

PROPOSED EXERCISES:

1. In `AvoidanceNodeNode`, instead of using the nearest obstacle, uses all nearby detected obstacles to compute the repulsion vector.

2. In `ObjectDetector`, instead of calculating a building block that encloses all the pixels that pass the filter, calculate a bounding box for each independent object. Publish the bounding boxes corresponding to the object most recently detected.

3. Try to make `HeadController` more reactive.

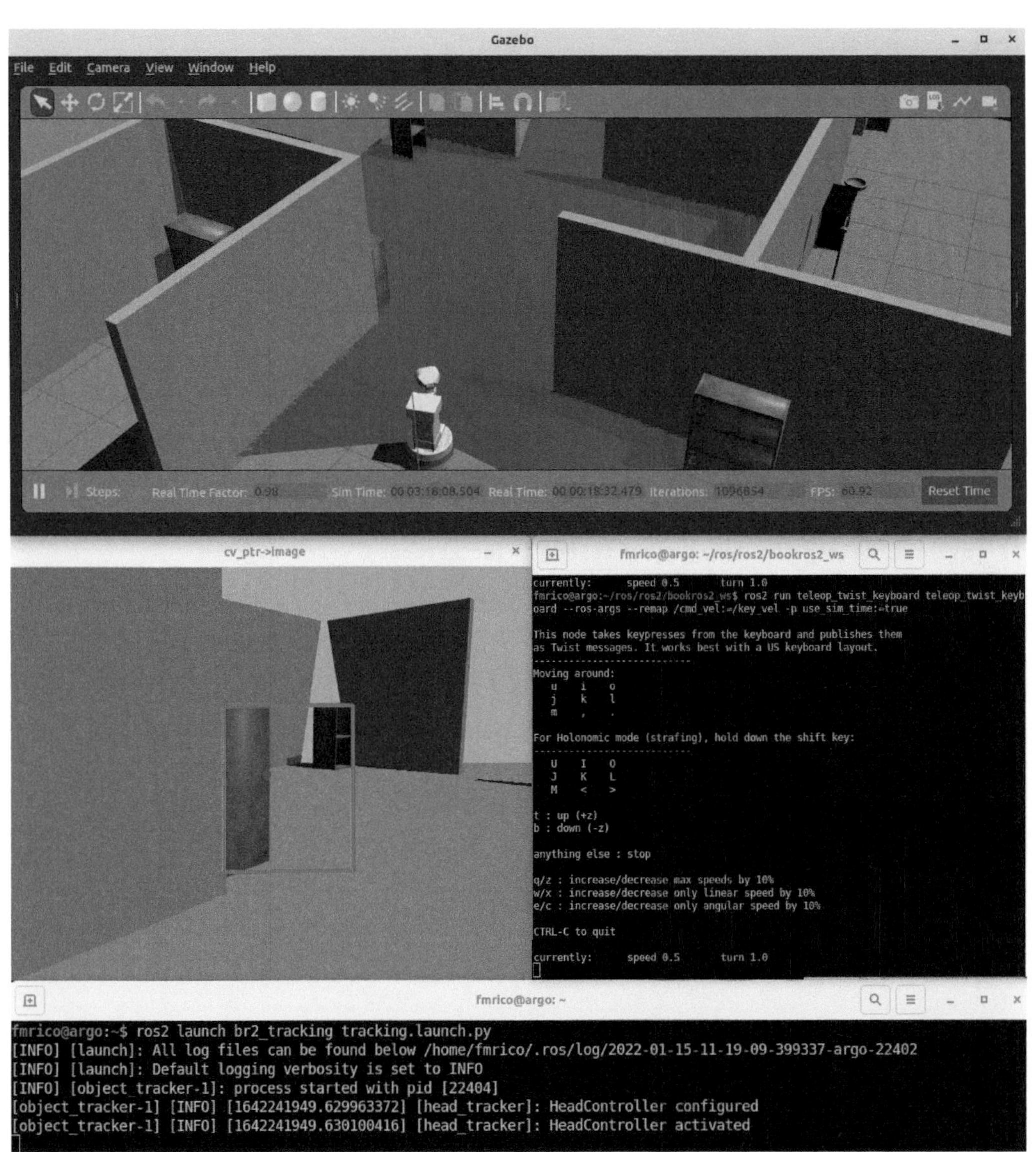

Figure 5.11: Project tracking running.

## CHAPTER 6

Programming Robot Behaviors with Behavior Trees

BEHAVIOR Trees for robot control [4] have become very popular in recent years. They have been used in various applications, mainly in video games and robots. They are usually compared to finite state machines, but the reality is that they are different approximations. When developing robotic behaviors with (Finite State Machines) FSMs, we think about states and transitions. When we use Behavior Trees, we think of sequences, fallbacks, and many flow resources that give them great expressiveness. In this chapter, as an illustrative example, we will implement the *Bump and Go* that we did with FSMs in the Chapter 3, and we will see how much the two approaches differ.

6.1 BEHAVIOR TREES

A Behavior Tree (BT) is a mathematical model to encode the control of a system. A BT is a way to structure the switching between different tasks in an autonomous agent, such as a robot or a virtual entity in a computer game. It is a hierarchical data structure defined recursively from a root node with several child nodes. Each child node, in turn, can have more children, and so on. Nodes that do not have children are usually called leaves of the tree.

The basic operation of a node is the **tick**. When a node is ticked, it can return three different values:

- **SUCCESS**: The node has completed its mission successfully.
- **FAILURE**: The node has failed in its mission.
- **RUNNING**: The node has not yet completed its mission.

A BT has four different types of nodes:

DOI: 10.1201/9781003289623-6

- **Control**: These types of nodes have 1-N children. Its function is to spread the tick to their children.
- **Decorators**: They are control nodes with only one child.
- **Action**: They are the leaves of the tree. The user must implement action nodes since they must generate the control required by the application.
- **Condition**: They are action nodes that cannot return RUNNING. In this case, the value SUCCESS is understood as the condition it encodes is met, and FAILURE if it is not.

Figure 6.1 shows a simple BT. When a BT is executed, the root node is ticked until it finishes executing, that is, until it returns SUCCESS or FAILURE.

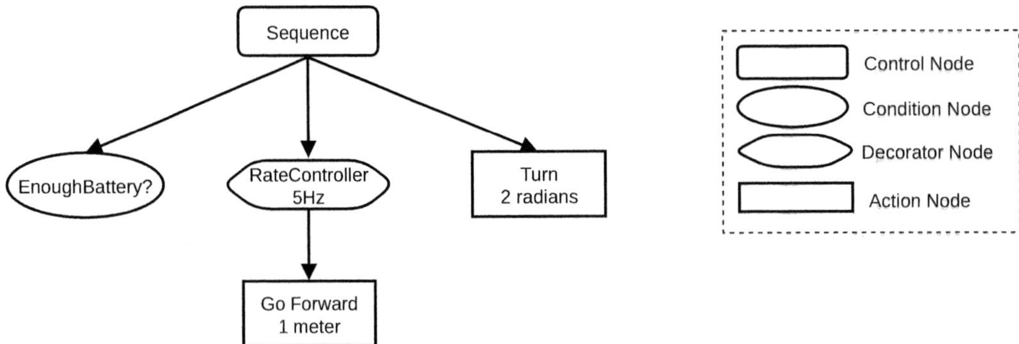

Figure 6.1: Simple Behavior Tree with various types of Nodes.

- The root node is a control node of type `Sequence`. This node ticks its children in order, starting from left. When a child returns SUCCESS, the sequence node ticks the next one. If the child node returns something else, the sequence node returns this value.
- The first child, `EnoughtBattery?`, is a Condition node. If it returns SUCCESS, it indicates that there is enough battery for the robot to carry out its mission so that the sequence node can advance to the next child. If it returned FAILURE, the mission would be aborted, as the result of executing the BT would be FAILURE.
- The `Go Forward` action node commands the robot to advance. As long as it has not traveled 1 m, the node returns RUNNING with each tick. When it has traveled the specified distance, it will return SUCCESS.
- The `Go Forward` action node has as its parent a Decorator node that controls that the frequency at which its child ticks is not greater than 5 Hz. Meanwhile, each tick returns the value returned by the child in the last tick.
- The `Turn` action node is similar to `Go Forward`, but spinning the robot 2 radians.

The library of available nodes can be extended with nodes created by the user. As we have said before, the user must implement the action nodes, but if we need any other type of node that is not available, we can implement it. In the above example, the `RateController` decorator node is not part of the Behavior Tree core library but can be implemented by the user.

A Behavior Tree controls the action decision flow. Leaves are not intended to implement complex algorithms or subsystems. The BT leaves should coordinate other subsystems in the robot. In ROS2, this is done by publishing or subscribing to topics or using ROS2 services/actions. Figure 6.2 shows a BT in which the nodes are used to coordinate the actions of a robot. Observe how the complexity is in the subsystem that coordinates, not in the BT leaves.

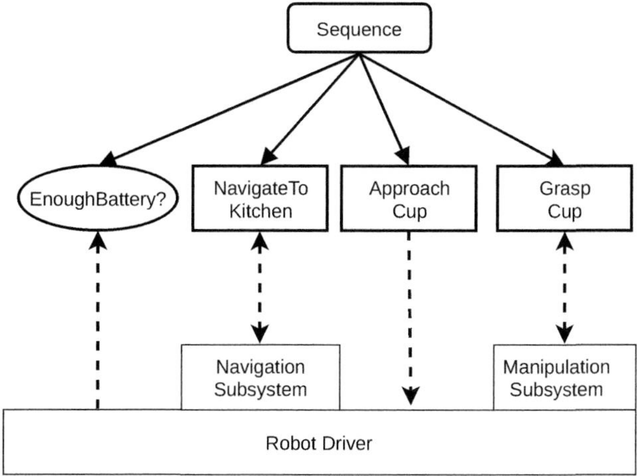

Figure 6.2: BT where the leaves control a robot by publish/subscribe (one-way dotted arrow) or ROS2 actions (two-way dotted arrow).

The second control node that we will present is the `Fallback`. This node can express fallback strategies, that is, what to do if a node returns FAILURE. Figure 6.3 shows an example of the use of this node.

1. The `Fallback` node ticks the first child. If it returns FAILURE, it ticks the next child.

2. If the second child returns SUCCESS, the `Fallback` node returns SUCCESS. Otherwise, it ticks the next child.

3. If all children have returned FAILURE, the `Fallback` node returns FAILURE.

In the development cycle with Behavior Trees, we can identify two phases:

- **Node Development**:Action nodes and any other node that the user requires for their application are designed, developed, and compiled in this phase. These nodes become part of the library of available nodes at the same category as the core nodes of Behavior Trees.

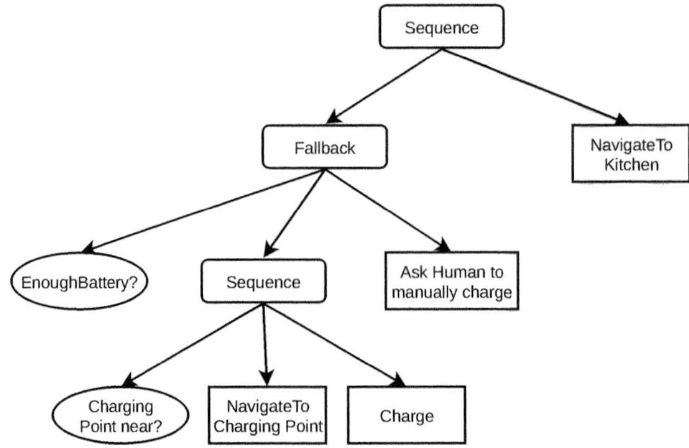

Figure 6.3: BT with a fallback strategy for charging battery.

- **Deployment**: In this phase, the Behavior Tree is composed using the available nodes. It is important to note that multiple different Behavior Trees can be created with the same nodes. If the nodes have been designed sufficiently generally, in this phase, very different behaviors of the robot can be defined using the same nodes.

A Behavior Tree has a *blackboard*, a key/value storage that all nodes in a tree can access. Nodes can have input ports and output ports to exchange information between them. The output ports of one node are connected to the input ports of another node using a key from the blackboard. While the ports of the nodes (their type and port class) have to be known at compile-time, the connections are established at deployment-time.

Figure 6.4 shows an example of connecting nodes through ports. A `DetectObject` action node is in charge of detecting some object so that the `InformHuman` node communicates it to the robot operator. `DetectObject` uses its output port `detected_id` to send the identifier of the detected object to `InformHuman` through its port `object_id`. For this, they use the input of the blackborad whose key `objID` currently has the value `cup`. Using keys from the blackboard is not mandatory. At deployment time, the value could be a constant value.

Behavior Trees are specified in XML. Although editing tools such as Groot[1] are used, they generate a BT in XML format. If this BT is saved to disk and this file is loaded from an application, any change to the BT does not require recompiling. The format is easy to understand, and it is widespread for BTs to be designed directly in XML. The following code shows two equally valid alternatives for the BT in Figure 6.1.

[1]https://github.com/BehaviorTree/Groot

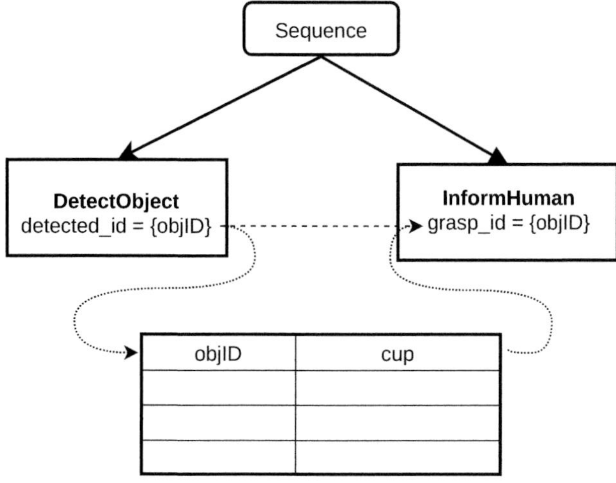

Figure 6.4: Ports connection using a blackboard key.

Compact XML syntax

```
<BehaviorTree ID="BehaviorTree">
    <Sequence>
        <EnoughBattery/>
        <RateController Rate="5Hz">
            <GoForward distance="1.0"/>
        </RateController>
        <Turn angle="2.0"/>
    </Sequence>
</BehaviorTree>
```

Extended XML syntax

```
<?xml version="1.0"?>
<root main_tree_to_execute="BehaviorTree">

    <BehaviorTree ID="BehaviorTree">
        <Sequence>
            <Condition ID="EnoughBattery"/>
            <Decorator ID="RateController" Rate="5Hz">
                <Action ID="GoForward" distance="1.0"/>
            </Decorator>
            <Action ID="Turn" angle="2.o"/>
        </Sequence>
    </BehaviorTree>

    <TreeNodesModel>
        <Condition ID="EnoughBattery"/>
        <Action ID="GoForward">
            <input_port name="distance"/>
        </Action>
        <Decorator ID="RateController">
            <input_port name="Rate"/>
        </Decorator>
        <Action ID="Turn">
            <input_port name="angle"/>
        </Action>
    </TreeNodesModel>
</root>
```

Table 6.1 shows a summary of the commonly available control nodes. This table shows what a control node returns when ticked, based on what the ticked child

Table 6.1: Summary of the behavior of the control nodes. Cell color groups into sequence, fallback, and decorator nodes.

	Value returned by child		
Control Node Type	**FAILURE**	**SUCCESS**	**RUNNING**
`Sequence`	Return FAILURE and restart sequence	Tick next child. Return SUCCESS if no more child	Return RUNNING and tick again
`ReactiveSequence`	Return FAILURE and restart sequence	Tick next child. Return SUCCESS if no more child	Return RUNNING and restart sequence
`SequenceStar`	Return FAILURE and tick again	Tick next child. Return SUCCESS if no more child	Return RUNNING and tick again
`Fallback`	Tick next child. Return FAILURE if no more child	Return SUCCESS	Return RUNNING and tick again
`ReactiveFallback`	Tick next child. Return FAILURE if no more child	Return SUCCESS	Return RUNNING and restart sequence
`InverterNode`	Return SUCCESS	Return FAILURE	Return RUNNING
`ForceSuccessNode`	Return SUCCESS	Return SUCCESS	Return RUNNING
`ForceFailureNode`	Return FAILURE	Return FAILURE	Return RUNNING
`RepeatNode` (N)	Return FAILURE	Return RUNNING N timesbefore returning SUCCESS	Return RUNNING
`RetryNode` (N)	Return RUNNING N times before returning FAILURE	Return SUCCESS	Return RUNNING

returns. In the case of sequences and fallbacks, it also shows what it does if this control node is ticked again: tick the next, restart the first child, or insist on the same child.

Let's analyze in detail some of these control nodes:

- **Sequence nodes**: In the previous section, we have used the basic sequence node. Behavior Trees allow sequence nodes with different behavior, which is helpful in some applications.
  - **`Sequence`**:As explained in the previous section, this node ticks its first child. When it returns SUCCESS, the ticks are made to the next child, and so on. If any child returns FAILURE, this node returns FAILURE and, if ticked again, starts over from the first child.

    Figure 6.5 shows an example of a sequence in which to take an image, it must check that the object is close and that the camera is ready. Once the camera is pointed at the subject, a picture can be taken. If any of the above children fail, the sequence fails. No child repeats its execution if it has already indicated that it has finished successfully.

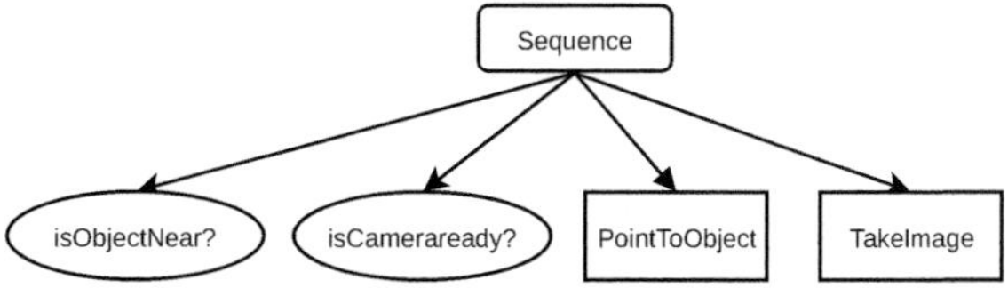

Figure 6.5: Example of Sequence node.

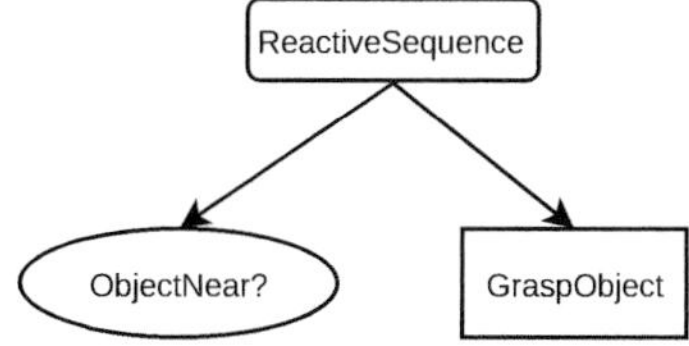

Figure 6.6: Example of ReactiveSequence node.

- **`ReactiveSequence`**: this sequence is commonly used when it is necessary to check conditions continuously. If any child returns RUNNING, the sequence restarts from the beginning. In this way, all nodes are always ticked up to the one returned by RUNNING on the previous tick.
- **`SequenceStar`**: This sequence is used to avoid restarting a sequence if, at some point, it has returned a FAILURE child. If this sequence is ticked again after a failure, the failed node is ticked directly.

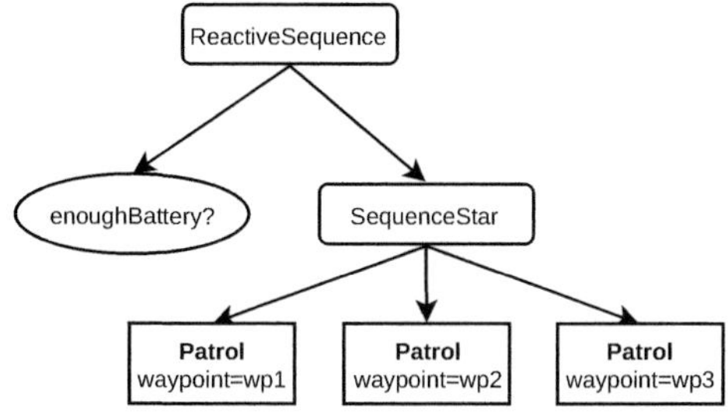

Figure 6.7: Example of ReactiveStar node.

- **Fallback nodes**: As we presented previously, fallback nodes allow us to execute different strategies to satisfy a condition until we find a successful one.
  - **`Fallback`**: It is the basic version of this control node. The children tick in sequence. When one returns FAILURE, it moves on to the next. The moment one returns SUCCESS, the fallback node returns SUCCESS.
  - **`ReactiveFallback`**: This alternative version of fallback has the difference that if a node returns RUNNING, the sequence is restarted from the beginning. The next tick will be made again to the first child. It is useful when the first node is a condition, and it must be checked while executing the action that tries to satisfy it. For example, in Figure 6.8, the action of charging the robot is running while the battery is not charged.

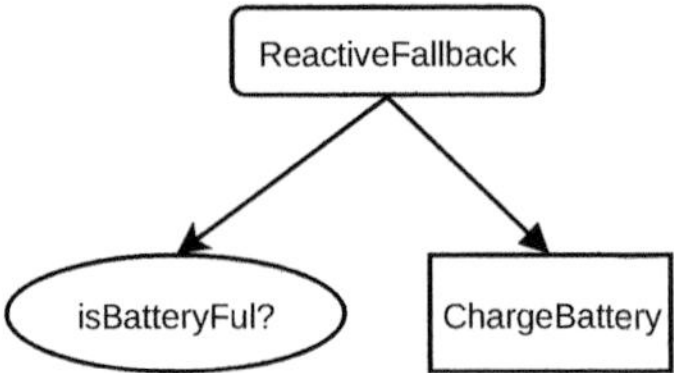

Figure 6.8: Example of ReactiveFallback node.

- **Decorator nodes**: They modify the return value of their only child. In the case of `RepeatNode` and `RetryNode`, they receive the `N` repetitions or retries through their input port.

6.2 *BUMP AND GO* WITH BEHAVIOR TREES

In this section we will show how to implement action nodes within our ROS2 packages, and how these nodes can access the Computation Graph to communicate with other nodes. To do this, we will reimplement the *Bump and Go* example that we did with state machines in Chapter 3, and thus we will see the differences that exist.

Let's start with the design of the Behavior Tree (Figure 6.10). It seems clear that we will need the following BT nodes (Figure 6.9):

- A condition node that indicates whether there is an obstacle (SUCCESS) or not (FAILURE) depending on the information received from the laser.
- Three action nodes that make the robot turn, move or go forward publishing speed messages. `Back` and `Turn` will return RUNNING for 3 s before returning SUCCESS. `Forward` will return RUNNING in all ticks.

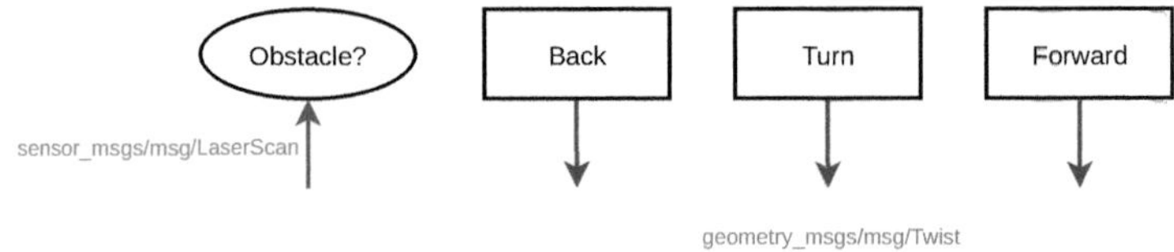

Figure 6.9: Action nodes for *Bump and Go*.

The Computation Graph is similar to the one in Figure 3.4, so we will skip its explanation. Let's focus on the workspace:

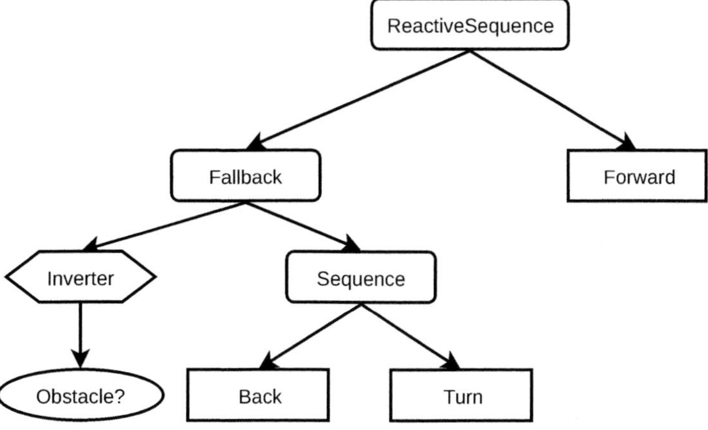

Figure 6.10: Complete Behavior Tree for *Bump and Go*.

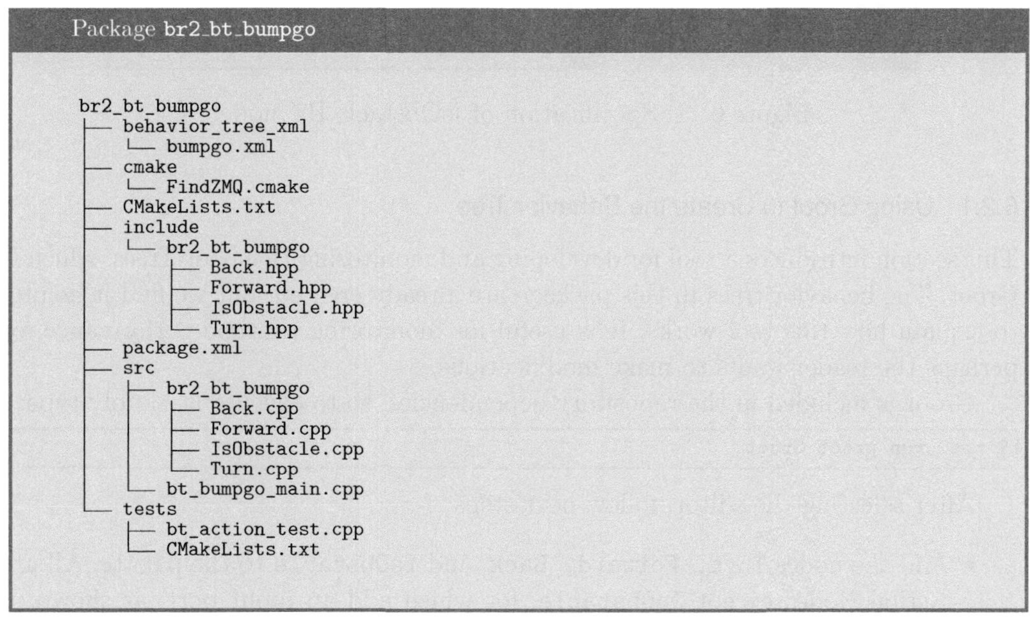

Package br2_bt_bumpgo

```
br2_bt_bumpgo
├── behavior_tree_xml
│   └── bumpgo.xml
├── cmake
│   └── FindZMQ.cmake
├── CMakeLists.txt
├── include
│   └── br2_bt_bumpgo
│       ├── Back.hpp
│       ├── Forward.hpp
│       ├── IsObstacle.hpp
│       └── Turn.hpp
├── package.xml
├── src
│   ├── br2_bt_bumpgo
│   │   ├── Back.cpp
│   │   ├── Forward.cpp
│   │   ├── IsObstacle.cpp
│   │   └── Turn.cpp
│   └── bt_bumpgo_main.cpp
└── tests
    ├── bt_action_test.cpp
    └── CMakeLists.txt
```

- Each of the BT nodes is a C++ class. Just like when we implement ROS2 nodes, we create a directory with the package name in `src` for sources and a directory with the same name in `include` for headers.
- A `tests` directory where there will be tests with gtest and a program to manually test a BT node, as we will explain later.
- A `cmake` directory contains a cmake file to find the ZMQ[2] library needed to debug Behavior Trees at runtime.
- A `behavior_tree_xml` directory with XML files containing the structure of the behavior trees that we will use in this package.

[2]https://zeromq.org/

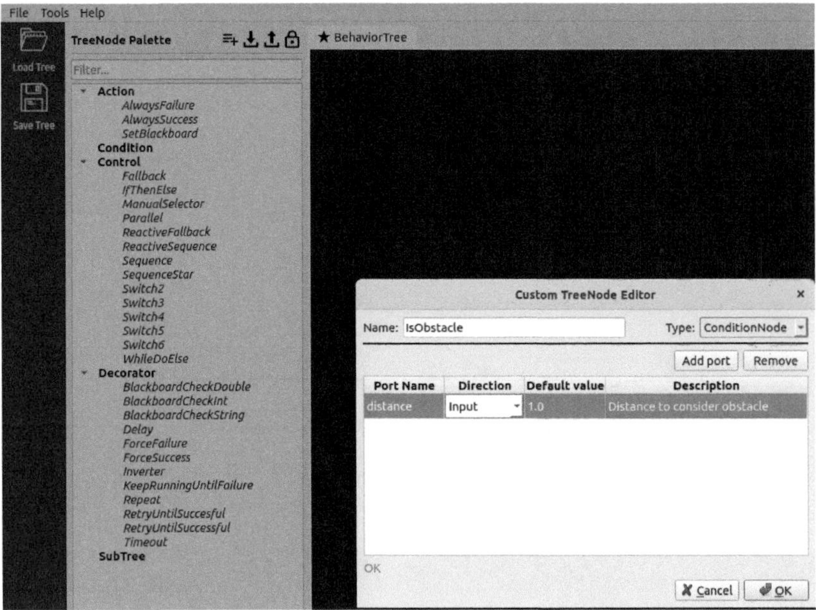

Figure 6.11: Specification of IsObstacle BT node.

6.2.1 Using Groot to Create the Behavior Tree

This section introduces a tool for developing and monitoring Behavior Trees, which is Groot. The behavior trees in this package are already created, but we find it helpful to explain how this tool works. It is useful for monitoring runtime performance, or perhaps the reader wants to make modifications.

Groot is included in the repository dependencies, so to execute it, simply type:

```
$ ros2 run groot Groot
```

After selecting the editor, follow next steps:

- Add the nodes `Turn, Forward, Back`, and `IsObstacle` to the palette. All are Action Nodes except `IsObstacle`, for which add an input port, as shown in Figure 6.11.
- Save the palette.
- Create the Behavior Tree as shown in Figure 6.12.
- Save the Behavior Tree in `mr2_bt_bumpgo/behavior_tree_xml/bumpgo.xml`.

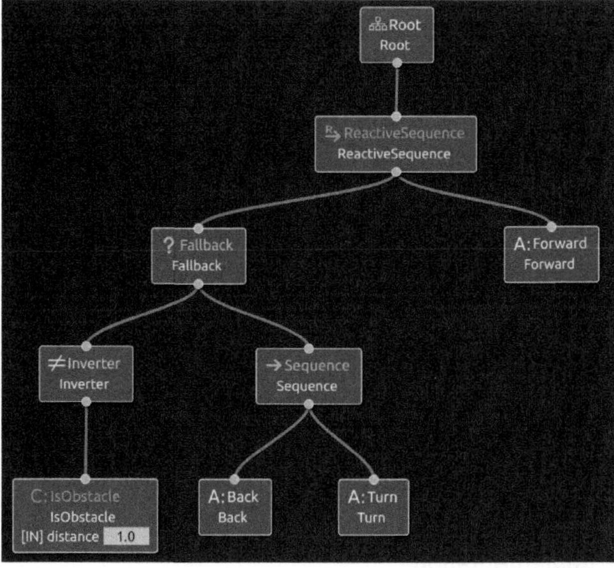

Figure 6.12: Action nodes for *Bump and Go*.

behavior_tree_xml/bumpgo.xml

```
<?xml version="1.0"?>
<root main_tree_to_execute="BehaviorTree">
    <BehaviorTree ID="BehaviorTree">
        <ReactiveSequence>
            <Fallback>
                <Inverter>
                    <Condition ID="IsObstacle" distance="1.0"/>
                </Inverter>
                <Sequence>
                    <Action ID="Back"/>
                    <Action ID="Turn"/>
                </Sequence>
            </Fallback>
            <Action ID="Forward"/>
        </ReactiveSequence>
    </BehaviorTree>
    <TreeNodesModel>
        <Action ID="Back"/>
        <Action ID="Forward"/>
        <Condition ID="IsObstacle">
            <input_port default="1.0" name="distance">Dist to consider obst</input_port>
        </Condition>
        <Action ID="Turn"/>
    </TreeNodesModel>
</root>
```

The Behavior Tree specification in XML is straightforward. There are two parts:

- **BehaviorTree**: It is the specification of the tree structure. The XML tags match the type of BT node specified, and the child nodes are within their parents.
- **TreeNodesModel**: Define the custom nodes we have created, indicating their input and output ports.

There is a valid alternative to this structure, which is ignoring the **TreeNodesModel** and directly using the name of the custom BT nodes:

```
<?xml version="1.0"?>
<root main_tree_to_execute="BehaviorTree">
    <BehaviorTree ID="BehaviorTree">
        <ReactiveSequence>
            <Fallback>
                <Inverter>
                    <IsObstacle distance="1.0"/>
                </Inverter>
                <Sequence>
                    <Back/>
                    <Turn/>
                </Sequence>
            </Fallback>
            <Forward/>
        </ReactiveSequence>
    </BehaviorTree>
</root>
```

6.2.2 BT Nodes Implementation

We will use the Behavior Trees library behaviortree.CPP[3], which is pretty standard in ROS/ROS2. Let's look at the `Forward` implementation to get an idea of how simple it is to implement a BT node:

include/mr2_bt_bumpgo/Forward.hpp

```
class Forward : public BT::ActionNodeBase
{
public:
  explicit Forward(
    const std::string & xml_tag_name,
    const BT::NodeConfiguration & conf);

  BT::NodeStatus tick();

  static BT::PortsList providedPorts()
  {
    return BT::PortsList({});
  }

private:
  rclcpp::Node::SharedPtr node_;
  rclcpp::Time start_time_;
  rclcpp::Publisher<geometry_msgs::msg::Twist>::SharedPtr vel_pub_;
};
```

As shown in the previous code, when a Behavior Tree is created, an instance of each class of a BT node is constructed for each one that appears in the Behavior Tree. An action node inherits from `BT::ActionNodeBase`, having to implement three methods and setting the constructor arguments:

- The constructor receives the content of the name field (which is optional) in the XML, as well as a `BT::NodeConfiguration` that contains, among other things, a pointer to the blackboard shared by all the nodes of a tree.

[3]https://www.behaviortree.dev/

- The `halt` method is called when the tree finishes its execution, and it is used to carry out any cleanup that the node requires. We will define void, as it is a pure virtual method.
- The `tick` method implements the tick operation that we have already described in this chapter.
- A static method that returns the ports of the node. In this case, `Forward` has no ports, so we return an empty list of ports.

The class definition is also straightforward:

src/mr2_bt_bumpgo/Forward.cpp

```
Forward::Forward(
  const std::string & xml_tag_name,
  const BT::NodeConfiguration & conf)
: BT::ActionNodeBase(xml_tag_name, conf)
{
  config().blackboard->get("node", node_);

  vel_pub_ = node_->create_publisher<geometry_msgs::msg::Twist>("/output_vel", 100);
}

BT::NodeStatus
Forward::tick()
{
  geometry_msgs::msg::Twist vel_msgs;
  vel_msgs.linear.x = 0.3;
  vel_pub_->publish(vel_msgs);

  return BT::NodeStatus::RUNNING;
}

}  // namespace br2_bt_bumpgo

#include "behaviortree_cpp_v3/bt_factory.h"
BT_REGISTER_NODES(factory)
{
  factory.registerNodeType<br2_bt_bumpgo::Forward>("Forward");
}
```

- In the constructor, after calling the constructor of the base class, we will get the pointer to the ROS2 node of the blackboard. We will see soon that when the tree is created, the pointer to the ROS2 node is inserted into the blackboard with the key "node" so that it is available to any BT node that requires it to create publishers, subscribers, get the time, or any related task to ROS2.
- The `tick` method is quite obvious: each time the node is ticked, it publishes a speed message to go forward, and return RUNNING.
- In the last part of the previous code, we register this class as implementing the `Forward` BT node. This part will be used when creating the tree.

Once the BT node `Forward` has been analyzed, the rest of the nodes are implemented similarly. Let's see some peculiarities:

- The BT node `Turn` performs its task for 3 s, so it saves the timestamp of its first tick, which is identifiable because its state is still IDLE:

src/mr2_bt_bumpgo/Turn.cpp

```
BT::NodeStatus
Turn::tick()
{
  if (status() == BT::NodeStatus::IDLE) {
    start_time_ = node_->now();
  }

  geometry_msgs::msg::Twist vel_msgs;
  vel_msgs.angular.z = 0.5;
  vel_pub_->publish(vel_msgs);

  auto elapsed = node_->now() - start_time_;

  if (elapsed < 3s) {
    return BT::NodeStatus::RUNNING;
  } else {
    return BT::NodeStatus::SUCCESS;
  }
}
```

- The BT node `isObstacle` saves the laser readings and compares them to the distance set on its input port:

src/mr2_bt_bumpgo/isObstacle.cpp

```
void
IsObstacle::laser_callback(sensor_msgs::msg::LaserScan::UniquePtr msg)
{
  last_scan_ = std::move(msg);
}

BT::NodeStatus
IsObstacle::tick()
{
  double distance = 1.0;
  getInput("distance", distance);

  if (last_scan_->ranges[last_scan_->ranges.size() / 2] < distance) {
    return BT::NodeStatus::SUCCESS;
  } else {
    return BT::NodeStatus::FAILURE;
  }
}
```

Each of the BT nodes will be compiled as a separate library. Later we will see that, when creating the Behavior Tree that contains them, we can load these libraries as plugins, quickly locating the implementation of the custom BT nodes.

CMakeLists.txt

```
add_library(br2_forward_bt_node SHARED src/br2_bt_bumpgo/Forward.cpp)
add_library(br2_back_bt_node SHARED src/br2_bt_bumpgo/Back.cpp)
add_library(br2_turn_bt_node SHARED src/br2_bt_bumpgo/Turn.cpp)
add_library(br2_is_obstacle_bt_node SHARED src/br2_bt_bumpgo/IsObstacle.cpp)

list(APPEND plugin_libs
  br2_forward_bt_node
  br2_back_bt_node
  br2_turn_bt_node
  br2_is_obstacle_bt_node
)

foreach(bt_plugin ${plugin_libs})
  ament_target_dependencies(${bt_plugin} ${dependencies})
  target_compile_definitions(${bt_plugin} PRIVATE BT_PLUGIN_EXPORT)
endforeach()

install(TARGETS
  ${plugin_libs}
  ARCHIVE DESTINATION lib
  LIBRARY DESTINATION lib
  RUNTIME DESTINATION lib/${PROJECT_NAME}
)
```

6.2.3 Running the Behavior Tree

Running a Behavior Tree is easy. A program should build a tree and start ticking its root until it returns SUCCESS. Behavior trees are created using a `BehaviorTreeFactory`, specifying an XML file or directly a string that contains the XML. `BehaviorTreeFactory` needs to load the libraries of the custom nodes as plugins and needs the blackboard to be shared among the BT nodes.

To integrate behavior trees with ROS2, create a ROS2 node and put it on the blackboard. As shown before, BT nodes can extract it from the blackboard to create publishers/subscribers or clients/servers of services or actions. Along with the tick at the root of the tree, a `spin_some` manages the arrival of messages to the ROS2 node.

See how it looks like the program that carries out the tree creation and execution:

src/mr2_bt_bumpgo/isObstacle.cpp

```
int main(int argc, char * argv[])
{
  rclcpp::init(argc, argv);

  auto node = rclcpp::Node::make_shared("patrolling_node");

  BT::BehaviorTreeFactory factory;
  BT::SharedLibrary loader;

  factory.registerFromPlugin(loader.getOSName("br2_forward_bt_node"));
  factory.registerFromPlugin(loader.getOSName("br2_back_bt_node"));
  factory.registerFromPlugin(loader.getOSName("br2_turn_bt_node"));
  factory.registerFromPlugin(loader.getOSName("br2_is_obstacle_bt_node"));

  std::string pkgpath = ament_index_cpp::get_package_share_directory("br2_bt_bumpgo");
  std::string xml_file = pkgpath + "/behavior_tree_xml/bumpgo.xml";
```

src/mr2_bt_bumpgo/isObstacle.cpp

```
  auto blackboard = BT::Blackboard::create();
  blackboard->set("node", node);
  BT::Tree tree = factory.createTreeFromFile(xml_file, blackboard);

  auto publisher_zmq = std::make_shared<BT::PublisherZMQ>(tree, 10, 1666, 1667);

  rclcpp::Rate rate(10);

  bool finish = false;
  while (!finish && rclcpp::ok()) {
    finish = tree.rootNode()->executeTick() != BT::NodeStatus::RUNNING;

    rclcpp::spin_some(node);
    rate.sleep();
  }

  rclcpp::shutdown();
  return 0;
}
```

1. At the beginning of the main function, we create a generic ROS2 node which we then insert into the blackboard. This is the node that we have seen that is pulled from the blackboard in `Forward` to create the speed message publisher.

2. The tree is created by a `BT::BehaviorTreeFactory` from an XML, BT action nodes that we will be implemented, and a blackboard.

   (a) As we will see below, each BT node will be compiled as an independent library. The `loader` object helps to find the library in the system to load the BT Node as a plugin. The `BT_REGISTER_NODES` macro that we saw earlier in the BT nodes definition allows the BT node name to be connected with its implementation within the library.

   ```
   BT::BehaviorTreeFactory factory;
   BT::SharedLibrary loader;

   factory.registerFromPlugin(loader.getOSName("br2_forward_bt_node"));
   ```

   (b) Function `get_package_share_directory` from package `ament_index_cpp` lets to obtain the full path of installed package, in order to read any file within. Remember that this is a package included in the package dependencies.

   ```
   std::string pkgpath = ament_index_cpp::get_package_share_directory(
     "br2_bt_bumpgo");
   std::string xml_file = pkgpath + "/behavior_tree_xml/forward.xml";
   ```

   (c) Finally, after creating the blackboard and inserting the shared pointer to the ROS2 node there, the factory builds the tree to execute.

   ```
   auto blackboard = BT::Blackboard::create();
   blackboard->set("node", node);
   BT::Tree tree = factory.createTreeFromFile(xml_file, blackboard);
   ```

(d) To debug the Behavior Tree at runtime, create `PublisherZMQ` object that publishes all the necessary information. To create it, indicate the tree, the maximum messages per second, and the network ports to use.

```
auto publisher_zmq = std::make_shared<BT::PublisherZMQ>(
  tree, 10, 1666, 1667);
```

3. In this last part, the tree's root is ticked at 10 Hz while the tree returns RUNNING while handling any pending work in the node, such as the delivery of messages that arrive at subscribers.

Once compiled, execute the simulator and the node and run the program. The robot should move forward.

```
$ ros2 launch br2_tiago sim.launch.py
```

```
$ ros2 run br2_bt_bumpgo bt_bumpgo --ros-args -r input_scan:=/scan_raw -r
output_vel:=/key_vel -p use_sim_time:=true
```

During program execution, it is possible to use Groot to monitor the state of the Behavior Tree to know which nodes are being ticked and the values they return. Simply boot up Groot and select Monitor instead of Editor. Once pressed connect, monitor the execution, as shown in Figure 6.13.

6.2.4 Testing the BT Nodes

Two types of tests have been included in this package that has been useful during this project's development. They are all in the tests directory of the package.

The first type of test has been to manually test each node separately, running behavior trees that only contain one type of node to see if they work correctly in isolation. We have included only the verification of the BT node Forward:

tests/bt_forward_main.cpp

```
factory.registerFromPlugin(loader.getOSName("br2_forward_bt_node"));

std::string xml_bt =
  R"(
  <root main_tree_to_execute = "MainTree" >
    <BehaviorTree ID="MainTree">
        <Forward />
    </BehaviorTree>
  </root>)";

auto blackboard = BT::Blackboard::create();
blackboard->set("node", node);
BT::Tree tree = factory.createTreeFromText(xml_bt, blackboard);

rclcpp::Rate rate(10);
bool finish = false;
while (!finish && rclcpp::ok()) {
  finish = tree.rootNode()->executeTick() != BT::NodeStatus::RUNNING;

  rclcpp::spin_some(node);
  rate.sleep();
}
```

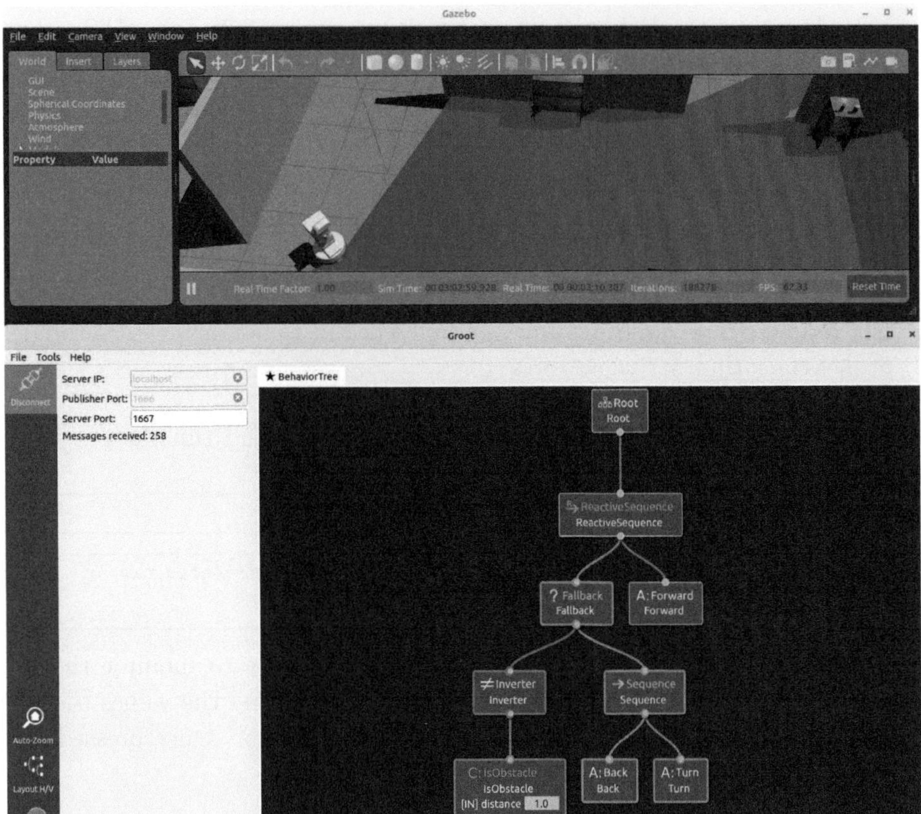

Figure 6.13: Monitoring the execution of a Behavior Tree with Groot.

Start the simulator and run:

```
$ build/br2_bt_bumpgo/tests/bt_forward --ros-args -r input_scan:=/scan_raw -r
output_vel:=/key_vel -p use_sim_time:=true
```

Check that the robot will go forward forever. Do the same with the rest of the BT nodes.

The second type of test is the one recommended in the previous chapter, which is using GoogleTest. It is easy to define a ROS2 node that records what speeds have been sent to the speed topic.

tests/bt_action_test.cpp

```
class VelocitySinkNode : public rclcpp::Node
{
public:
  VelocitySinkNode()
  : Node("VelocitySink")
  {
    vel_sub_ = create_subscription<geometry_msgs::msg::Twist>(
      "/output_vel", 100, std::bind(&VelocitySinkNode::vel_callback, this, _1));
  }

  void vel_callback(geometry_msgs::msg::Twist::SharedPtr msg)
  {
    vel_msgs_.push_back(*msg);
  }

  std::list<geometry_msgs::msg::Twist> vel_msgs_;

private:
  rclcpp::Subscription<geometry_msgs::msg::Twist>::SharedPtr vel_sub_;
};
```

It is possible to execute a tree for a few cycles, checking that the speeds that were sent were correct:

tests/bt_action_test.cpp

```
TEST(bt_action, forward_btn)
{
  auto node = rclcpp::Node::make_shared("forward_btn_node");
  auto node_sink = std::make_shared<VelocitySinkNode>();

  // Creation the Behavior Tree only with the Forward BT node

  rclcpp::Rate rate(10);
  auto current_status = BT::NodeStatus::FAILURE;
  int counter = 0;
  while (counter++ < 30 && rclcpp::ok()) {
    current_status = tree.rootNode()->executeTick();
    rclcpp::spin_some(node_sink);
    rate.sleep();
  }

  ASSERT_EQ(current_status, BT::NodeStatus::RUNNING);
  ASSERT_FALSE(node_sink->vel_msgs_.empty());
  ASSERT_NEAR(node_sink->vel_msgs_.size(), 30, 1);

  geometry_msgs::msg::Twist & one_twist = node_sink->vel_msgs_.front();

  ASSERT_GT(one_twist.linear.x, 0.1);
  ASSERT_NEAR(one_twist.angular.z, 0.0, 0.0000001);
}
```

In this case, after ticking the root of the tree 30 times, see how the node is still returning RUNNING, 30 speed messages have been advertised, and the speeds are correct (they move the robot forward). We could have examined all of them, but we have only done this case for the first one.

Examine the tests of the other nodes. In the case of `Turn` and `Back`, it is checked that they do so for the appropriate time before returning success. In the case of isObstacle, we create synthetic laser readings to see if the output is correct in all cases.

6.3 PATROLLING WITH BEHAVIOR TREES

In this section, we will address a more complex and ambitious project. We have previously said that Behavior Tree action nodes help control other subsystems. In the project of the previous section, we have done it in a pretty basic way, processing sensory information and sending speeds. In this section, we will carry out a project in which a Behavior Tree will control more complex subsystems, such as the Nav2 Navigation subsystem and the active vision subsystem that we developed in the previous chapter.

The goal of the project in this section is that of a robot patrolling the simulated house in Gazebo:

- The robot patrols three waypoints in the house (Figure 6.14). Upon reaching each waypoint, the robot turns on itself for a few seconds to perceive its surroundings.
- While the robot goes from one waypoint to another, the robot perceives and tracks the detected objects.
- The robot keeps track (simulated) of the battery level it has. When low, it goes to a recharge point to recharge for a few seconds.

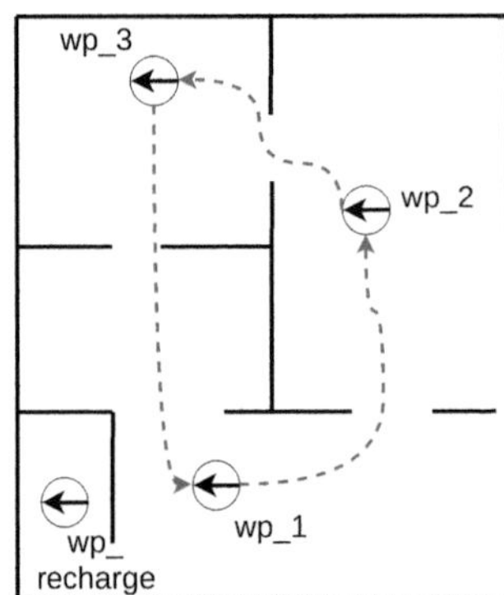

Figure 6.14: Waypoints at the simulated home, with the path followed during patrolling.

Since we are using such a complex and important subsystem as Nav2, the navigation system in ROS2, we will first describe it in Section 6.3.1. The Section 6.3.2 describes the steps to set up Nav2 for a particular robot and environment. It is possible to skip this section since the `br2_navigation` package already contains the environment map and configuration files for the simulated Tiago scenario in the house. The following sections already focus on implementing the Behavior Tree and the patrolling nodes.

6.3.1 Nav2 Description

Nav2[4][3] is the ROS2 navigation system designed to be modular, configurable, and scalable. Like its predecessor in ROS, it aspires to be the most widely used navigation software, so it supports major robot types: holonomic, differential-drive, legged, and Ackermann (car-like) while allowing information from lasers and 3D cameras to be merged, among others. Nav2 incorporates multiple plugins for local and global navigation and allows custom plugins to be easily used.

The inputs to Nav2 are TF transformations (conforming to REP-105), a map[5], any relevant sensor data sources. It also requires the navigation logic, coded as a BT XML file coded, adapting it to specific problems if needed. Nav2 outputs are the speed sent to the robot base.

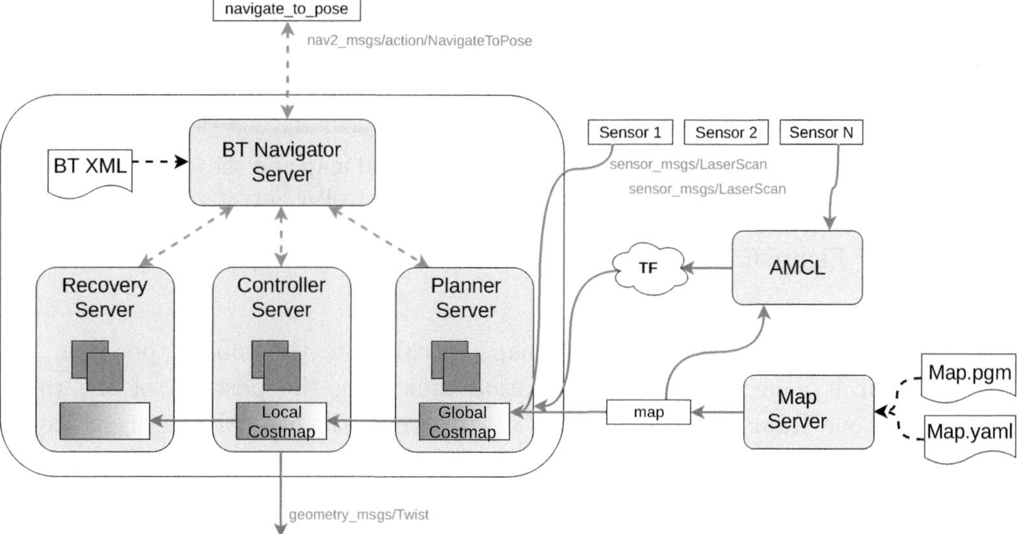

Figure 6.15: Waypoints at the simulated home, with the path followed during patrolling.

Nav2 has the modular architecture shown in Figure 6.15. Let's describe what each of the components that appear in the figure are:

- **Map Server**: This component reads a map from two files and publishes it as a `nav_msgs/msg/OccupancyGrid`, which nodes internally handle as a `costmap2D`. The maps in Nav2 are grids whose cells encode whether the space is free (0), unknown (255), or occupied (254). Values between 1 and 253 encode different occupation degrees or cost to cross this area. Figure 6.16b shown the map coded as a `costmap2D`.
- **AMCL**: This component implements a localization algorithm based on Adaptive Monte-Carlo (AMCL) [9]. It uses sensory information, primarily distance

[4]https://navigation.ros.org/

[5]if using the Static Costmap Layer

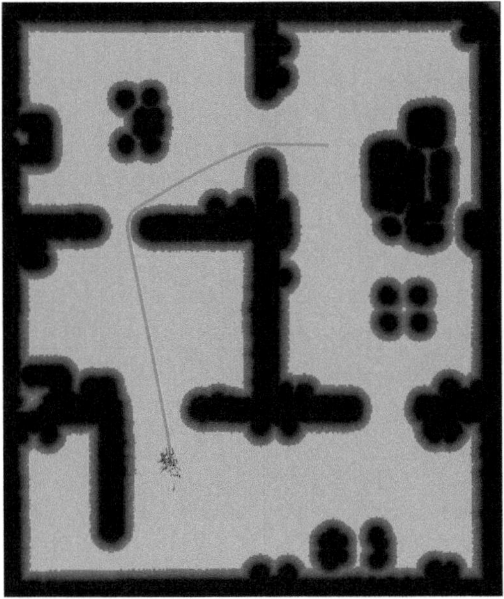

(a) Global costmap used by the Planner Server.

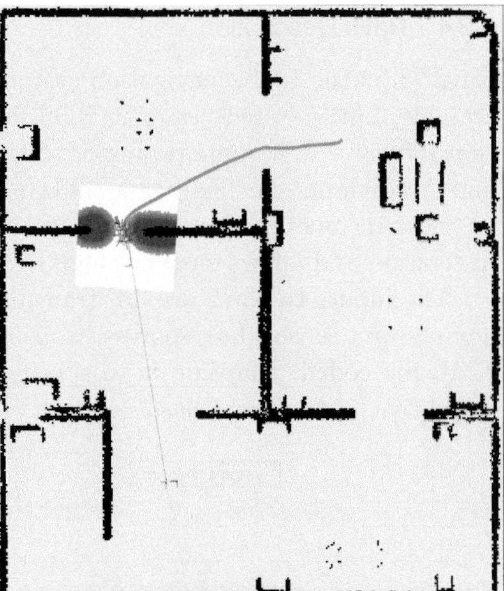

(b) Original map and the local costmap used by the Controller Server.

Figure 6.16: 2D costmaps used by the Nav2 components.

readings from a laser and the map, to calculate the robot's position. The output is a geometric transformation indicating the position of the robot. Since one frames should not have two parents, instead of posting a `map` → `base_footprint` transform, this component computes and publishes the `map` → `odom` transform.

- **Planner Server**: The function of this component is to calculate a route from the origin to the destination. It takes as input the destination, the current position of the robot, and the map of the environment. The Planner Server builds a costmap from the original map, whose walls are fattened with the radius of the robot and a certain safety margin. The idea is that the robot uses the free space (or with low cost) to calculate the routes, as shown in Figure 6.16a. Route planning and costmap updating algorithms are loaded as plugins. Like the following two components, this component receives requests through ROS2 actions.

- **Controller Server**: This component receives the route calculated by the Planner Server and publishes the speeds sent to the robot base. It uses a costmap of the robot's surroundings (see Figure 6.16b), where nearby obstacles are encoded and used by algorithms (loaded as plugins) to calculate speeds.

- **Recovery Server**: This component has several helpful recovery strategies if the robot gets lost, gets stuck, or cannot calculate routes to the destination. These strategies are turning, clearing costmaps, slow-moving, among others.

- **BT Navigator Server**: This is the component that orchestrates the rest of the navigation components. It receives navigation requests in the form of ROS2 actions. The action name is `navigate_to_pose` and the type is `nav2_msgs/action/NavigateToPose`. Therefore, if we want to make a robot go from one point to another, we must use this ROS2 action. Check out what this action looks like:

```
ros2 interface show nav2_msgs/action/NavigateToPose

#goal definition
geometry_msgs/PoseStamped pose
string behavior_tree
---
#result definition
std_msgs/Empty result
---
geometry_msgs/PoseStamped current_pose
builtin_interfaces/Duration navigation_time
int16 number_of_recoveries
float32 distance_remaining
```

  - The request section comprises a target position and, optionally, a custom Behavior Tree to be used in this action instead of the default one. This last feature allows special requests to be made that are not normal navigation behavior, such as following a moving object or approaching an obstacle in a particular way.
  - The result of the action, when finished.
  - The robot continuously returns the current position and the distance to the target and statistical data such as the navigation time or the times it has recovered from undesirable situations.

  BT Navigator uses Behavior Trees to orchestrate robot navigation. The Behavior Tree nodes make requests to the other components of Nav2 so that they carry out their task.

  When this component accepts a navigation action, it starts executing a Behavior Tree like the one shown in Figure 6.17. Nav2's default Behavior Tree is quite a bit more complex, including calls to recoveries, but the one in the figure is quite illustrative of BT Navigator Server's use of them. First, the goal that arrives in the ROS2 action is put on the blackboard. ComputePathToPose uses this goal to call the Planner Server action, which returns a route to it. This path is the output of this BT node which is input to the BT node `FollowPath`, which sends it to the Controller Server.

To use Nav2, it is enough to install the packages that contain it:

```
$ sudo apt install ros-foxy-navigation2 ros-foxy-nav2-bringup
ros-foxy-turtlebot3*
```

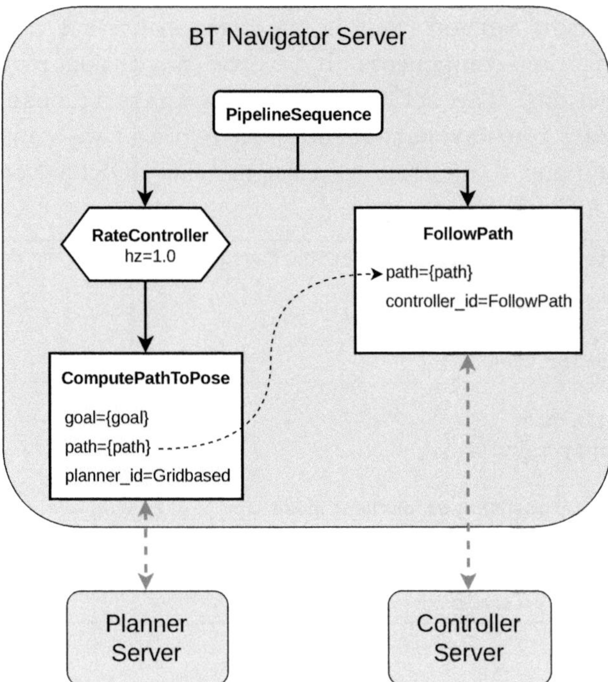

Figure 6.17: Behavior Tree simple example inside BT Navigator Server, with BT nodes calling ROS2 actions to coordinate other Nav2 components.

In the `br2_navigation` package, we have prepared the necessary launchers, maps, and configuration files for the simulated Tiago robot to navigate in the home scenario. Let's test navigation:

1. Launch the simulator:

```
$ ros2 launch br2_tiago sim.launch.py
```

2. Launch navigation:

```
$ ros2 launch br2_navigation tiago_navigation.launch.py
```

3. Open RViz2 and display (see Figure 6.18):

    - TF: To display the robot. Observe the transformation `map` $\rightarrow$ `odom`.
    - Map: Display the topic `/map`, which QoS is reliable and transient local.
    - Global Costmap: Display the topic `/global_costmap/costmap` with default QoS (Reliable and Volatile).
    - Local Costmap: Display the topic `/local_costmap/costmap` with default QoS.
    - LaserScan: To see how it matches with obstacles.

- It is interesting to display the AMCL particles. Each one is a hypothesis about the robot's position. The final robot position is the mean of all these particles. As much concentrated is this population of arrows, better localized is the robot. It is in the `/particlecloud` with which QoS is best effort + volatile.

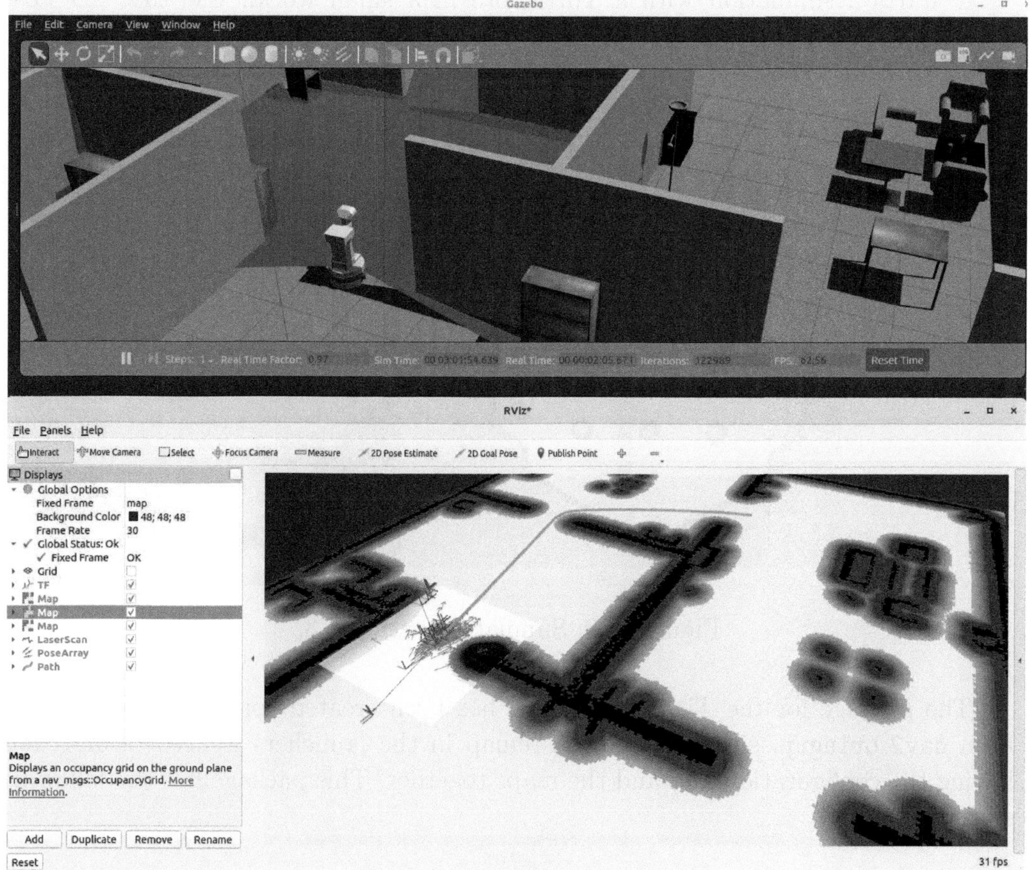

Figure 6.18: Nav2 in action

4. Use the "2D Goal Pose" button to command a goal position to the robot.

5. In obtaining a map position, use the "Publish Point" button. Then click in any position on the map. This position will be published in the topic `/clicked_point`.

6.3.2 Setup Nav2

This section describes the Nav2 setup process for a new environment and a specific robot. It is possible to can skip it, as the `br2_navigation` package already contains everything you need to make the simulated Tiago navigate in the house scenario. Keep reading for using another scenario or another robot.

If Nav2 is installed from packages, it is in `/opt/ros/foxy/`. In particular, in `/opt/ros/foxy/share/nav2_bringup` is the Nav2 bringup package with launchers, maps, and parameters for a simulated Turtlebot3[6] that comes by default and that you can launch by typing:

```
$ ros2 launch nav2_bringup tb3_simulation_launch.py
```

It starts a simulation with a Turtlebot3 in a small world. Use the "2D Pose Estimate" button to put where the robot is (see Figure 6.19), as the navigation will not be activated until then.

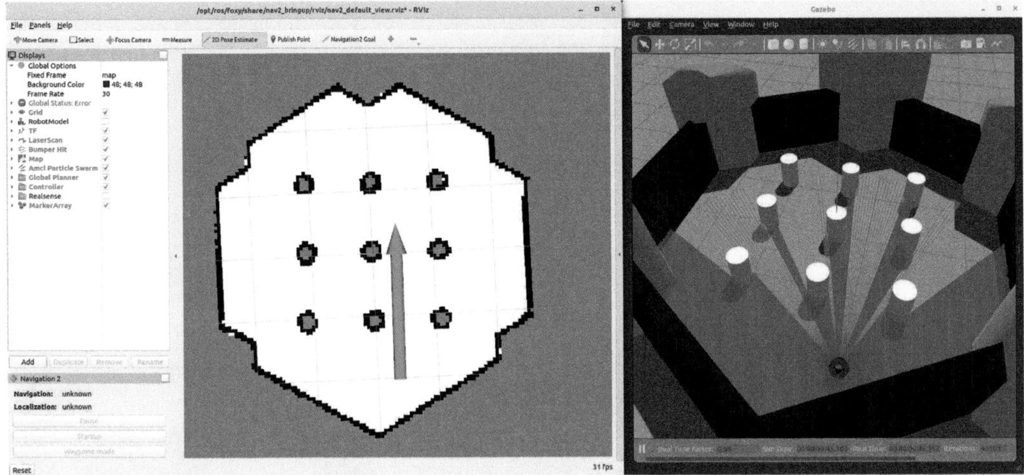

Figure 6.19: Simulated turtlebot 3.

The package for the Tiago simulation has been created copying some elements from `nav2_bringup`, since some extra remap in the launchers is needed, and thus having the configuration files and the maps together. This package has the following structure:

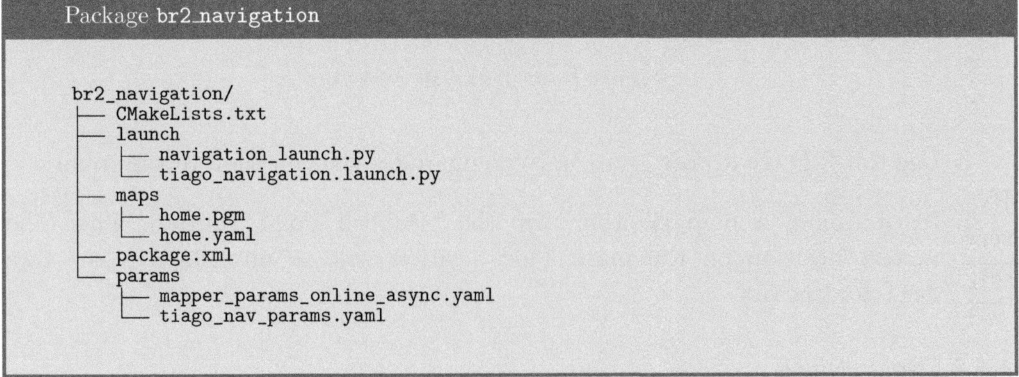

Package `br2_navigation`

```
br2_navigation/
├── CMakeLists.txt
├── launch
│   ├── navigation_launch.py
│   └── tiago_navigation.launch.py
├── maps
│   ├── home.pgm
│   └── home.yaml
├── package.xml
└── params
    ├── mapper_params_online_async.yaml
    └── tiago_nav_params.yaml
```

Start by looking at how to map the environment. We will use the `slam_toolbox` package. We will use a custom param file to specify the particular topics and frames:

[6]https://emanual.robotis.com/docs/en/platform/turtlebot3/overview/

params/mapper_params_online_async.yaml

```
# ROS Parameters
odom_frame: odom
map_frame: map
base_frame: base_footprint
scan_topic: /scan_raw
mode: mapping #localization
```

Run these commands, each in a different terminal:

1. Tiago simulated with the home scenario.

```
$ ros2 launch br2_tiago sim.launch.py
```

2. RViz2 to visualize the mapping progress (see Figure 6.20).

```
$ rviz2 --ros-args -p use_sim_time:=true
```

3. Launch the SLAM node. It will publish in /map the map as far as it is being built.

```
$ ros2 launch slam_toolbox online_async_launch.py params_file:=[Full path
to bookros2_ws/src/book_ros2/br2_navigation/params/mapper_params_online_async
.yaml] use_sim_time:=true
```

4. Launch the map saver server. This node will subscribe to /map, and it will save it to disk when requested.

```
$ ros2 launch nav2_map_server map_saver_server.launch.py
```

5. Run the teleoperator to move the robot along the scenario.

6.
```
$ ros2 run teleop_twist_keyboard teleop_twist_keyboard --ros-args --remap
/cmd_vel:=/key_vel -p use_sim_time:=true
```

Run these commands, each in a different terminal:

As soon as the robot starts moving around the stage using the teleoperator, run RViz2 and check how the map is built. When the map is completed, ask the map server saver to save the map to disk:

```
$ ros2 run nav2_map_server map_saver_cli --ros-args -p use_sim_time:=true
```

Note that when mapping/navigating with a real robot, the use_sim_time parameters, both in launchers and nodes, must be false.

At this point, two files will have been created. A PGM image file (which you can modify if you need to do any fix) and a YAML file containing enough information to interpret the image as a map. Remember that if modifying the name of the files, this YAML should be modified too:

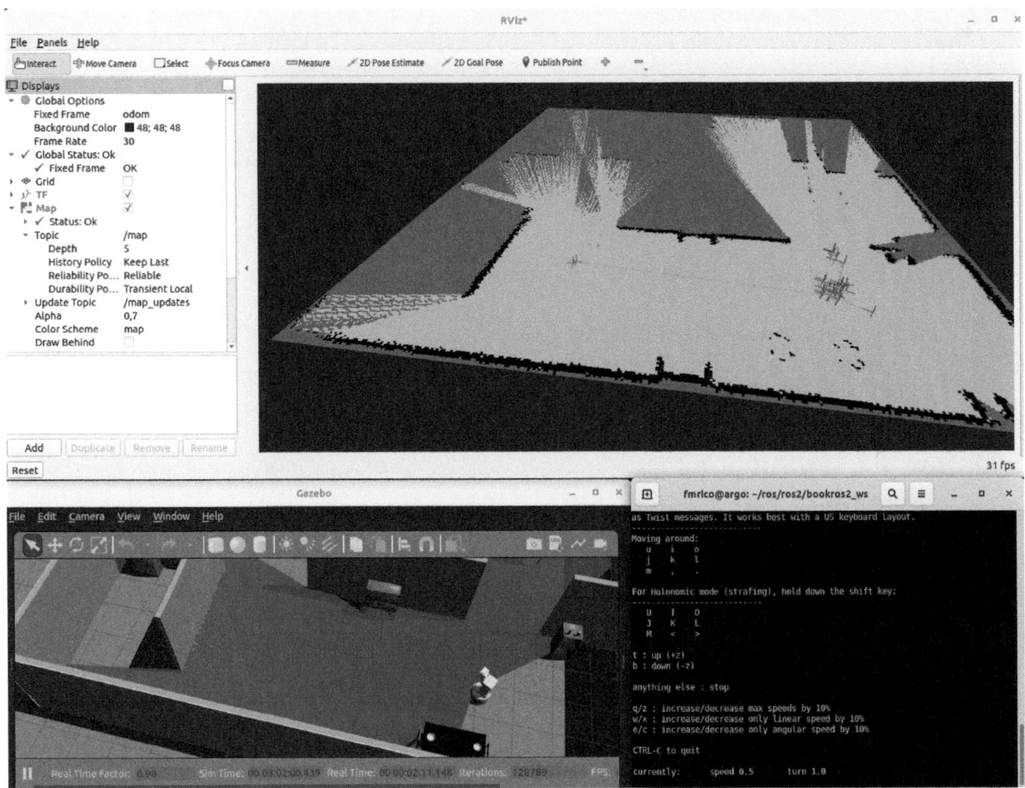

Figure 6.20: SLAM with Tiago simulated.

```
image: home.pgm
mode: trinary
resolution: 0.05
origin: [-2.46, -13.9, 0]
negate: 0
occupied_thresh: 0.65
free_thresh: 0.25
```

Move this file to the `br2_navigation` package and continue to the next setup step. In this step, the launchers copied from `nav2_bringup` needs to be modified. `tiago_navigation.launch` launch navigation and localization by including their launchers. We don't use directly the launchers in `nav2_bringup` because some extra remaps in `navigation.launch` has to be done.

br2_navigation/launch/navigation_launch.py

```
remappings = [('/tf', 'tf'),
('/tf_static', 'tf_static'),
('/cmd_vel', '/nav_vel')
]
```

Regarding the parameter files, start from the ones in the package `nav2_bringup`. Let's see some details on the configuration:

- First, and most important, set all the parameters that contain a sensor topic to the correct ones, and ensure that all the frames exist in our robot and are correct.
- If the initial position is known, set it in the AMCL configuration. If you start the robot in the same pose as you started when mapping, this is the $(0, 0, 0)$ position.

br2_navigation/params/tiago_nav_params

```
amcl:
  ros__parameters:
    scan_topic: scan_raw
    set_initial_pose: true
    initial_pose:
    x: 0.0
    y: 0.0
    z: 0.0
    yaw: 0.0
```

- Set the speeds and acceleration depending on the robot's capabilities:

br2_navigation/params/tiago_nav_params

```
controller_server:
  ros__parameters:
    use_sim_time: False
    FollowPath:
      plugin: "dwb_core::DWBLocalPlanner"
      min_vel_x: 0.0
      min_vel_y: 0.0
      max_vel_x: 0.3
      max_vel_y: 0.0
      max_vel_theta: 0.5
      min_speed_xy: 0.0
      max_speed_xy: 0.5
      min_speed_theta: 0.0
      acc_lim_x: 1.5
      acc_lim_y: 0.0
      acc_lim_theta: 2.2
      decel_lim_x: -2.5
      decel_lim_y: 0.0
      decel_lim_theta: -3.2
```

- Set the robot radius to inflate walls and obstacles and a scaling factor in setting how far navigate from them. These settings are held by the `inflation_layer` costmap plugin, applicable to local and global costmap:

br2_navigation/params/tiago_nav_params

```
local_costmap:
  local_costmap:
    ros__parameters:
      robot_radius: 0.3
      plugins: ["voxel_layer", "inflation_layer"]
      inflation_layer:
        plugin: "nav2_costmap_2d::InflationLayer"
        cost_scaling_factor: 3.0
        inflation_radius: 0.55
```

6.3.3 Computation Graph and Behavior Tree

The Computation Graph (Figure 6.21) of this project is made up of the node `patrolling_node` and the nodes that belong to the two subsystems that are being controlled: Nav2 and the active vision system developed in the last chapter.

- Nav2 is controlled using ROS2 actions, sending the goal poses that make up the patrol route.
- Regarding the active vision system during navigation, the HeadController node, a LifeCycleNode, will be activated (using ROS2 services).
- Also, upon arrival at a waypoint, to make the robot rotate on its own, `patrolling_node` will post velocities directly to the robot's base.

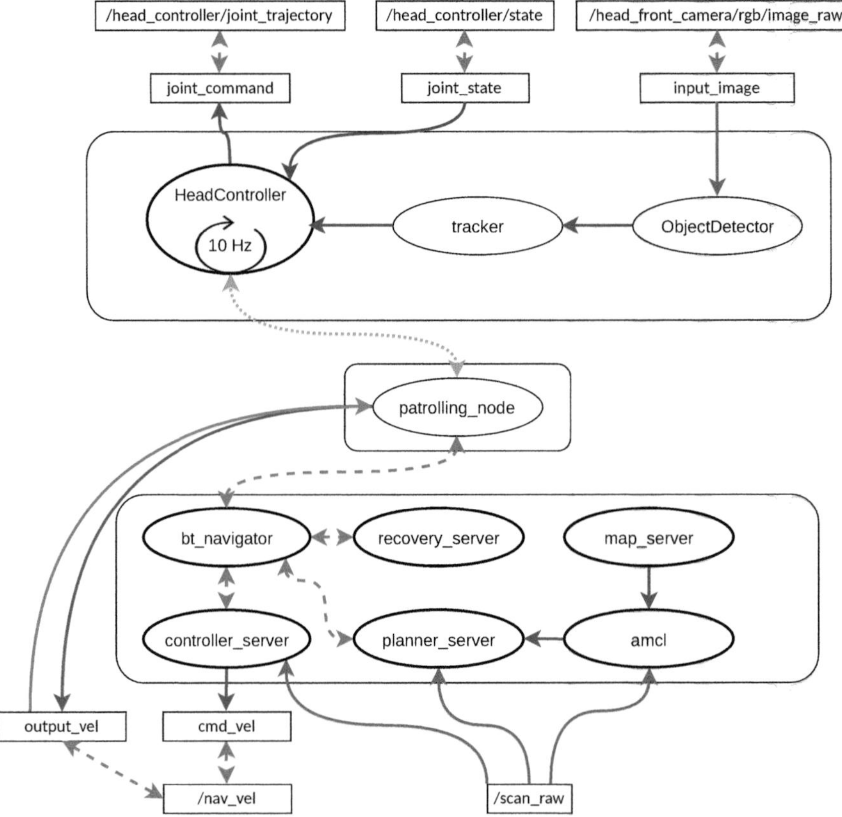

Figure 6.21: Computation Graph for the Patrolling Project. Subsystems have been simplified for clarity.

The `patrolling_node` node in the Computation Graph is shown to be quite simple. Perhaps it is more interesting to analyze the Behavior Tree that it contains, which is the one that controls its control logic. Figure 6.22 shows its complete structure. Analyze each one of its action and condition nodes:

- **Move**: This node is in charge of sending a navigation request to Nav2 through a ROS2 action. The navigation goal is received through an input port, in its goal port, which is a coordinate that contains an (x, y) position and a *theta* orientation. This node returns RUNNING until it is informed that the navigation action is complete, in which case it returns SUCCESS. The case in which it returns FAILURE has not been contemplated, although it would have been convenient.

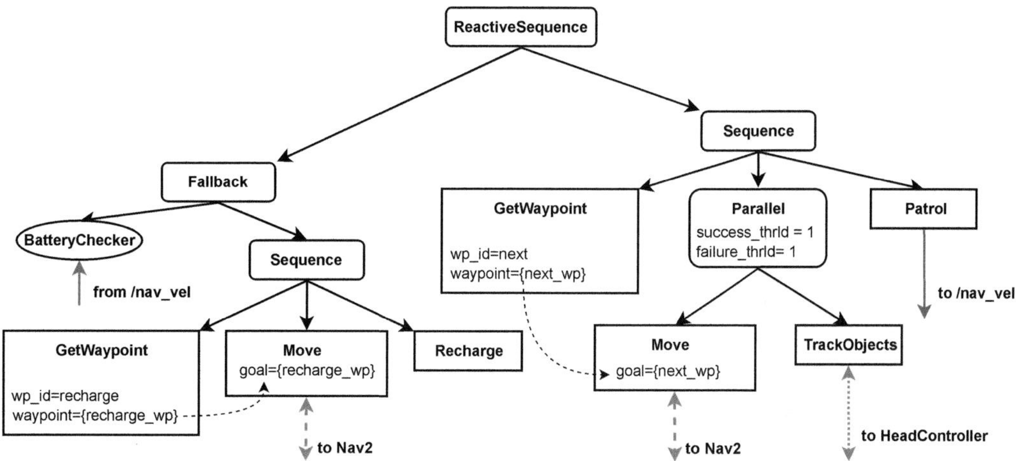

Figure 6.22: Behavior Tree for Patrolling project.

- **GetWaypoint**: This node is used to obtain the geometric coordinates used by Move. It has a waypoint output port with the geometric coordinates, which are then used by the BT node as input. The GetWayPoint input is an id indicating which waypoint is desired. If this input is "recharge", its output is the coordinates of the recharge point. If the input is "next", it returns the geometric coordinates of the next waypoint to navigate.

  This node exists because it simplifies the Behavior Tree since otherwise, the right branch of the tree would have to be repeated three times, once for each waypoint. The second is to delegate to another BT node the choice of the target point and thus simplify Move, not needing to maintain the coordinates of all the waypoints internally. There are many more alternatives, but this one is pretty clean and scalable.

- **BatteryChecker**: This node simulates the battery level of the robot. It keeps the battery level on the blackboard, decreasing over time and with the robot's movement (that is why it subscribes to the topic of commanded speeds). If the battery level drops below a certain level, it returns FAILURE. If not, return SUCCESS.

- **Patrol**: This node simply spins the robot around for a few seconds to control the environment. When it has finished, it returns SUCCESS.

- **TrackObjects**: This node always returns RUNNING. When it is first ticked, it activates, if it was not already, the `HeadController` node. This node runs in parallel with `Move`. The `Parallel` control node is configured so that when one of the two (and it can only be Move) returns SUCCESS, it considers that the task of all its children has finished, halting the nodes whose status is still RUNNING. When `TrackObjects` receives a halt, it disables the `HeadController`.

6.3.4 Patrolling Implementation

The structure of the `br2_bt_patrolling` package is similar to the one in the previous section: it implements each BT node separately, in the usual places for class definitions and declarations. It has a main program that creates the tree and executes it, and it has some tests for each of the implemented BT nodes.

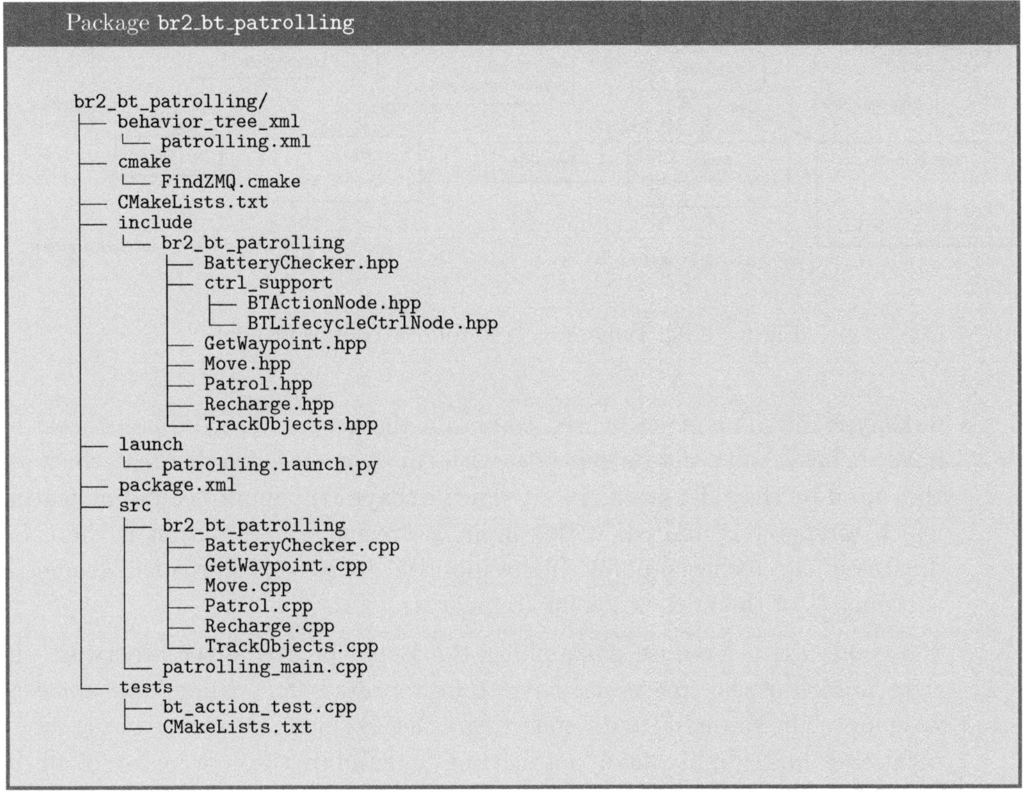

Package `br2_bt_patrolling`

```
br2_bt_patrolling/
├── behavior_tree_xml
│   └── patrolling.xml
├── cmake
│   └── FindZMQ.cmake
├── CMakeLists.txt
├── include
│   └── br2_bt_patrolling
│       ├── BatteryChecker.hpp
│       ├── ctrl_support
│       │   ├── BTActionNode.hpp
│       │   └── BTLifecycleCtrlNode.hpp
│       ├── GetWaypoint.hpp
│       ├── Move.hpp
│       ├── Patrol.hpp
│       ├── Recharge.hpp
│       └── TrackObjects.hpp
├── launch
│   └── patrolling.launch.py
├── package.xml
├── src
│   ├── br2_bt_patrolling
│   │   ├── BatteryChecker.cpp
│   │   ├── GetWaypoint.cpp
│   │   ├── Move.cpp
│   │   ├── Patrol.cpp
│   │   ├── Recharge.cpp
│   │   └── TrackObjects.cpp
│   └── patrolling_main.cpp
└── tests
    ├── bt_action_test.cpp
    └── CMakeLists.txt
```

From an implementation point of view, the most interesting are two classes that simplify the BT nodes that use ROS2 actions and those that activate a LifeCycleNode. They are in `include/br2_bt_patrolling/ctrl_support`, and have been implemented in a general way so that they can be reused for other projects.

The `BTActionNode` class has been borrowed from Nav2, where the BT Navigator Server used it to control the rest of its servers. It is quite a complex class since it considers many more cases than we use in this project, such as cancellation and resends of actions. We do not want to go into details about its implementation. I recommend the ROS2 actions tutorial on the official ROS2 page to learn more about ROS2 actions. When completed, come back to this class to explore this class.

BT nodes that wish to control a subsystem with ROS2 actions inherit this class. Let's analyze its interface to its derived class. Original comments will help us to understand their utility:

include/br2_bt_patrolling/ctrl_support/BTActionNode.hpp

```
template<class ActionT, class NodeT = rclcpp::Node>
class BtActionNode : public BT::ActionNodeBase
{
public:
  BtActionNode(
    const std::string & xml_tag_name,
    const std::string & action_name,
    const BT::NodeConfiguration & conf)
  : BT::ActionNodeBase(xml_tag_name, conf), action_name_(action_name)
  {
    node_ = config().blackboard->get<typename NodeT::SharedPtr>("node");
    ...
  }

  // Could do dynamic checks, such as getting updates to values on the blackboard
  virtual void on_tick()
  {
  }

  // Called upon successful completion of the action. A derived class can override this
  // method to put a value on the blackboard, for example.
  virtual BT::NodeStatus on_success()
  {
    return BT::NodeStatus::SUCCESS;
  }

  // Called when a the action is aborted. By default, the node will return FAILURE.
  // The user may override it to return another value, instead.
  virtual BT::NodeStatus on_aborted()
  {
    return BT::NodeStatus::FAILURE;
  }

  // The main override required by a BT action
  BT::NodeStatus tick() override
  {
    ...
  }

  // The other (optional) override required by a BT action. In this case, we
  // make sure to cancel the ROS2 action if it is still running.
  void halt() override
  {
    ...
  }
protected:
  typename ActionT::Goal goal_;
};
```

It is a template class because each action has a different type. In the case of `Move`, the action type is `nav2_msgs/action/NavigateToPose`. The class is also parameterized with the ROS2 node type because it may also be instantiated with LifeCycleNodes.

The `tick` and `halt` methods are handled by class `BtActionNode`, so they should not be defined in the derived class. The other methods can be overridden in the derived class to do something, like notifying when the action completes or fails. The derived class overrides `on_tick`, which is called once at startup, to set the goal. Let's see the implementation of `Move` inheriting from `BtActionNode`:

include/br2_bt_patrolling/Move.hpp

```
class Move : public br2_bt_patrolling::BtActionNode<nav2_msgs::action::NavigateToPose>
{
public:
  explicit Move(
    const std::string & xml_tag_name,
    const std::string & action_name,
    const BT::NodeConfiguration & conf);

  void on_tick() override;
  BT::NodeStatus on_success() override;

  static BT::PortsList providedPorts()
  {
    return {
      BT::InputPort<geometry_msgs::msg::PoseStamped>("goal")
    };
  }
};
```

src/br2_bt_patrolling/Move.cpp

```
Move::Move(
  const std::string & xml_tag_name,
  const std::string & action_name,
  const BT::NodeConfiguration & conf)
: br2_bt_patrolling::BtActionNode<nav2_msgs::action::NavigateToPose>(xml_tag_name,
    action_name, conf)
{
}

void
Move::on_tick()
{
  geometry_msgs::msg::PoseStamped goal;
  getInput("goal", goal);

  goal_.pose = goal;
}

BT::NodeStatus
Move::on_success()
{
  RCLCPP_INFO(node_->get_logger(), "navigation Suceeded");

  return BT::NodeStatus::SUCCESS;
}

#include "behaviortree_cpp_v3/bt_factory.h"
BT_REGISTER_NODES(factory)
{
  BT::NodeBuilder builder =
    [](const std::string & name, const BT::NodeConfiguration & config)
    {
      return std::make_unique<br2_bt_patrolling::Move>(
        name, "navigate_to_pose", config);
    };

  factory.registerBuilder<br2_bt_patrolling::Move>(
    "Move", builder);
}
```

- The BT Node `Move` implements `on_success` to report that the navigation has finished.
- The `on_tick` method gets the goal from the input port and assigns it to `goal_`. This variable will be the one that will be sent directly to Nav2.

- When building this BT node, the second argument is the name of the ROS2 action. In the case of Nav2, it is `navigate_to_pose`.

`BTLifecycleCtrlNode` is a class from which a BT Node is derived to activate/deactivate LifeCycle nodes. It is created specifying the name of the node to control. In the case of the `HeadTracker`, it will be with `/head_tracker`. All LifeCycleNodes have various services to be managed. In this, we will be interested in two:

- **`[node name]/get_state`**: Returns the status of a LifeCycleNode.
- **`[node name]/set_state`**: Sets the state of a LifeCycleNode.

Let's see code snippets of the `BTLifecycleCtrlNode` implementation:

include/br2_bt_patrolling/ctrl_support/BTLifecycleCtrlNode.hpp

```
class BtLifecycleCtrlNode : public BT::ActionNodeBase
{
public:
  BtLifecycleCtrlNode(...)
  : BT::ActionNodeBase(xml_tag_name, conf), ctrl_node_name_(node_name)
  {
  }

  template<typename serviceT>
  typename rclcpp::Client<serviceT>::SharedPtr createServiceClient(
    const std::string & service_name)
  {
    auto srv = node_->create_client<serviceT>(service_name);
    while (!srv->wait_for_service(1s)) {
      ...
    }
    return srv;
  }

  BT::NodeStatus tick() override
  {
    if (status() == BT::NodeStatus::IDLE) {
      change_state_client_ = createServiceClient<lifecycle_msgs::srv::ChangeState>(
        ctrl_node_name_ + "/change_state");
      get_state_client_ = createServiceClient<lifecycle_msgs::srv::GetState>(
        ctrl_node_name_ + "/get_state");
    }

    if (ctrl_node_state_ != lifecycle_msgs::msg::State::PRIMARY_STATE_ACTIVE) {
      ctrl_node_state_ = get_state();
      set_state(lifecycle_msgs::msg::State::PRIMARY_STATE_ACTIVE);
    }

    return BT::NodeStatus::RUNNING;
  }

  void halt() override
  {
    if (ctrl_node_state_ == lifecycle_msgs::msg::State::PRIMARY_STATE_ACTIVE) {
      set_state(lifecycle_msgs::msg::State::PRIMARY_STATE_INACTIVE);
    }
  }

  // Get the state of the controlled node
  uint8_t get_state(){...}

  // Get the state of the controlled node. Ot can fail, if not transition possible
  bool set_state(uint8_t state) {...}

  std::string ctrl_node_name_;
  uint8_t ctrl_node_state_;
};
```

Two clients are instantiated: one to query the state and one to set the state. They will be used in get_state and set_state, respectively. When the node is first ticked, the controlled node is requested to go to the active state. When halted, its deactivation is requested.

The BT node TrackObjects only needs to inherit from this class by specifying the name of the node:

include/br2_bt_patrolling/TrackObjects.hpp

```
class TrackObjects : public br2_bt_patrolling::BtLifecycleCtrlNode
{
public:
  explicit TrackObjects(
    const std::string & xml_tag_name,
    const std::string & node_name,
    const BT::NodeConfiguration & conf);

  static BT::PortsList providedPorts()
  {
    return BT::PortsList({});
  }
};
```

src/br2_bt_patrolling/TrackObjects.cpp

```
TrackObjects::TrackObjects(...)
: br2_bt_patrolling::BtLifecycleCtrlNode(xml_tag_name, action_name, conf)
{
}

#include "behaviortree_cpp_v3/bt_factory.h"
BT_REGISTER_NODES(factory)
{
  BT::NodeBuilder builder =
    [](const std::string & name, const BT::NodeConfiguration & config)
    {
      return std::make_unique<br2_bt_patrolling::TrackObjects>(
        name, "/head_tracker", config);
    };

  factory.registerBuilder<br2_bt_patrolling::TrackObjects>(
    "TrackObjects", builder);
}
```

Take into account that TrackObjects always returns RUNNING. That is why we have used it as a child of a parallel control node.

Check out how the rest of the BT nodes functionality has been implemented:

- **BatteryChecker**: The first difference between this BT node and the others is that it is a condition node. It does not have a halt method and cannot return RUNNING.

  This node checks the battery level stored on the blackboard at each tick. If it is less than a certain level, it returns FAILURE.

src/br2_bt_patrolling/BatteryChecker.cpp

```
const float MIN_LEVEL = 10.0;

BT::NodeStatus
BatteryChecker::tick()
{
  update_battery();

  float battery_level;
  config().blackboard->get("battery_level", battery_level);

  if (battery_level < MIN_LEVEL) {
    return BT::NodeStatus::FAILURE;
  } else {
    return BT::NodeStatus::SUCCESS;
  }
}
```

The `update_battery` method takes the battery level from the blackboard and decreases it in the function of time and the total amount of speed (`last_twist_`) currently requested. It is just a simulation of battery consumption.

src/br2_bt_patrolling/BatteryChecker.cpp

```
const float DECAY_LEVEL = 0.5;  // 0.5 * |vel| * dt
const float EPSILON = 0.01;  // 0.01 * dt

void
BatteryChecker::update_battery()
{
  float battery_level;
  if (!config().blackboard->get("battery_level", battery_level)) {
    battery_level = 100.0f;
  }

  float dt = (node_->now() - last_reading_time_).seconds();
  last_reading_time_ = node_->now();

  float vel = sqrt(last_twist_.linear.x * last_twist_.linear.x +
    last_twist_.angular.z * last_twist_.angular.z);
  battery_level = std::max(0.0f, battery_level - (vel * dt * DECAY_LEVEL) -
    EPSILON * dt);

  config().blackboard->set("battery_level", battery_level);
}
```

It is always useful to control de range of some calculus using `std::max` and `std::min`. In this case, we control that `battery_level` is never negative.

- **Recharge**: This BT node is related to the previous one. It takes some time to recharge the battery. Note that using blackboard lets some nodes collaborate to update and test some values.

src/br2_bt_patrolling/BatteryChecker.cpp

```
BT::NodeStatus
Recharge::tick()
{
  if (counter_++ < 50) {
    return BT::NodeStatus::RUNNING;
  } else {
    counter_ = 0;
    config().blackboard->set<float>("battery_level", 100.0f);
    return BT::NodeStatus::SUCCESS;
  }
}
```

Each BT node in the tree is a different instance of the same class, but be ready to be ticked even if the BT node returned once SUCCESS. In this case, restarting `counter_` to 0.

- **Patrol**: This node just makes the robot spin for 15 s. The only interesting aspect of this node is how it controls how long executing since the first tick until the it return SUCCESS. Take into account that a node status is IDLE the first tick, so it's possible to store this timestamp.

src/br2_bt_patrolling/Patrol.cpp

```
BT::NodeStatus
Patrol::tick()
{
  if (status() == BT::NodeStatus::IDLE) {
    start_time_ = node_->now();
  }

  geometry_msgs::msg::Twist vel_msgs;
  vel_msgs.angular.z = 0.5;
  vel_pub_->publish(vel_msgs);

  auto elapsed = node_->now() - start_time_;

  if (elapsed < 15s) {
    return BT::NodeStatus::RUNNING;
  } else {
    return BT::NodeStatus::SUCCESS;
  }
}
```

- **`GetWaypoint`**: This node stores the waypoint coordinates. If the input port `wp_id` is the string "recharge", its output is a coordinate, in frame `map`, corresponding to the position where it is supposed to be the robot charger. In another case, each time it is ticked, it returns the coordinates of a different waypoint.

src/br2_bt_patrolling/GetWaypoint.cpp

```
GetWaypoint::GetWaypoint(...)
{
  geometry_msgs::msg::PoseStamped wp;
  wp.header.frame_id = "map";
  wp.pose.orientation.w = 1.0;

  // recharge wp
  wp.pose.position.x = 3.67;
  wp.pose.position.y = -0.24;
  recharge_point_ = wp;

  // wp1
  wp.pose.position.x = 1.07;
  wp.pose.position.y = -12.38;
  waypoints_.push_back(wp);

  // wp2
  wp.pose.position.x = -5.32;
  wp.pose.position.y = -8.85;
  waypoints_.push_back(wp);
}
```

src/br2_bt_patrolling/GetWaypoint.cpp

```
BT::NodeStatus
GetWaypoint::tick()
{
  std::string id;
  getInput("wp_id", id);

  if (id == "recharge") {
    setOutput("waypoint", recharge_point_);
  } else {
    setOutput("waypoint", waypoints_[current_++]);
    current_ = current_ % waypoints_.size();
  }

  return BT::NodeStatus::SUCCESS;
}
```

6.3.5 Running Patrolling

From an implementation point of view, the only relevant thing is that we will use a launcher for the active vision system and the patrolling node. Navigation and the simulator could have been included in the launcher, but they generate so much output on the screen that we run them manually in other terminals. The launcher looks like this:

br2_navigation/launch/patrolling_launch.py

```
def generate_launch_description():
    tracking_dir = get_package_share_directory('br2_tracking')

    tracking_cmd = IncludeLaunchDescription(
        PythonLaunchDescriptionSource(os.path.join(tracking_dir, 'launch',
        'tracking.launch.py')))

    patrolling_cmd = Node(
        package='br2_bt_patrolling',
        executable='patrolling_main',
        parameters=[{
          'use_sim_time': True
        }],
        remappings=[
          ('input_scan', '/scan_raw'),
          ('output_vel', '/nav_vel')
        ],
        output='screen'
    )

    ld = LaunchDescription()
    ld.add_action(tracking_cmd)
    ld.add_action(patrolling_cmd)

    return ld
```

So type these commands, each one in a separate terminal:

```
$ ros2 launch br2_tiago sim.launch.py
```

```
$ ros2 launch br2_navigation tiago_navigation.launch.py
```

Nav2 also uses Behavior Trees and activates a server to debug its operation with Groot. It does this on Groot's default ports (1666 and 1667). For this reason, we have started it in 2666 and 2667. If we put them on the same, the program would fail.

Before connecting to the patrolling Behavior Tree, correctly set the ports to 2666 and 2667.

src/patrolling_main.cpp

```
BT::Tree tree = factory.createTreeFromFile(xml_file, blackboard);
auto publisher_zmq = std::make_shared<BT::PublisherZMQ>(tree, 10, 2666, 2667);
```

At this point, optionally open RViz2 to monitor navigation or Groot to monitor Behavior Tree execution. For the latter, wait to launch the patrol program to connect to the Behavior Tree:

```
$ rviz2 --ros-args -p use_sim_time:=true
```

Try sending a navigation position to make sure the navigation starts correctly.

```
$ ros2 run groot Groot
```

Finally, launch the patrol program together with the active vision system:

```
$ ros2 launch br2_bt_patrolling patrolling.launch.py
```

If everything has gone well, the robot, after recharging its battery, patrols the three waypoints established in the environment. While patrolling, observe how the robot tracks the objects it detects. When it reaches a waypoint and turns around, notice how tracking is no longer active. After a while of operation, the robot will run out of battery again, going to the recharging point again before continuing patrolling.

PROPOSED EXERCISES:

1. Make a program using Behavior Trees that makes the robot move continuously to the space without obstacles.

2. Explore the Nav2 functionality:

    - Mark forbidden areas in the center of each room in which the robot should not enter.
    - Modify the Behavior tree inside BT Navigator to finish navigation always one meter before the goal.
    - Try different Controller/Planner algorithms.

3. Publish the detected objects while patrolling as a 3D bounding box. You could do it by:

    - Using the pointcloud.
    - Using the depth image and the CameraInfo information. Like is done in:

      `https://github.com/gentlebots/gb_perception/blob/main/gb_perception_utils/src/gb_perception_utils/Perceptor3D.cpp`

Source Code

Complete list of source code used in the book, also available in:

https://github.com/fmrico/book_ros2

A.1 PACKAGE BR2_BASICS

Package br2_basics

```
br2_basics
├── CMakeLists.txt
├── config
│   └── params.yaml
├── launch
│   ├── includer_launch.py
│   ├── param_node_v1_launch.py
│   ├── param_node_v2_launch.py
│   ├── pub_sub_v1_launch.py
│   └── pub_sub_v2_launch.py
├── package.xml
└── src
    ├── executors.cpp
    ├── logger_class.cpp
    ├── logger.cpp
    ├── param_reader.cpp
    ├── publisher_class.cpp
    ├── publisher.cpp
    └── subscriber_class.cpp
```

DOI: 10.1201/9781003289623-A

br2_basics/CMakeLists.txt

```
cmake_minimum_required(VERSION 3.5)
project(br2_basics)

find_package(ament_cmake REQUIRED)
find_package(rclcpp REQUIRED)
find_package(std_msgs REQUIRED)

set(dependencies
    rclcpp
    std_msgs
)

add_executable(publisher src/publisher.cpp)
ament_target_dependencies(publisher ${dependencies})

add_executable(publisher_class src/publisher_class.cpp)
ament_target_dependencies(publisher_class ${dependencies})

add_executable(subscriber_class src/subscriber_class.cpp)
ament_target_dependencies(subscriber_class ${dependencies})

add_executable(executors src/executors.cpp)
ament_target_dependencies(executors ${dependencies})

add_executable(logger src/logger.cpp)
ament_target_dependencies(logger ${dependencies})

add_executable(logger_class src/logger_class.cpp)
ament_target_dependencies(logger_class ${dependencies})

add_executable(param_reader src/param_reader.cpp)
ament_target_dependencies(param_reader ${dependencies})


install(TARGETS
  publisher
  publisher_class
  subscriber_class
  executors
  logger
  logger_class
  param_reader
  ARCHIVE DESTINATION lib
  LIBRARY DESTINATION lib
  RUNTIME DESTINATION lib/${PROJECT_NAME}
)

install(DIRECTORY launch config DESTINATION share/${PROJECT_NAME})

if(BUILD_TESTING)
  find_package(ament_lint_auto REQUIRED)
  ament_lint_auto_find_test_dependencies()
endif()

ament_export_dependencies(${dependencies})
ament_package()
```

br2_basics/launch/param_node_v2_launch.py

```
import os

from ament_index_python.packages import get_package_share_directory

from launch import LaunchDescription
from launch_ros.actions import Node

def generate_launch_description():

  pkg_dir = get_package_share_directory('basics')
  param_file = os.path.join(pkg_dir, 'config', 'params.yaml')

  param_reader_cmd = Node(
    package='basics',
    executable='param_reader',
    parameters=[param_file],
    output='screen'
  )

  ld = LaunchDescription()
  ld.add_action(param_reader_cmd)

  return ld
```

br2_basics/launch/pub_sub_v2_launch.py

```
from launch import LaunchDescription
from launch_ros.actions import Node

def generate_launch_description():

  return LaunchDescription([
    Node(
      package='basics',
      executable='publisher',
      output='screen'
    ),
    Node(
      package='basics',
      executable='subscriber_class',
      output='screen'
    )
  ])
```

br2_basics/launch/pub_sub_v1_launch.py

```
from launch import LaunchDescription
from launch_ros.actions import Node

def generate_launch_description():

  pub_cmd = Node(
    package='basics',
    executable='publisher',
    output='screen'
  )
  sub_cmd = Node(
    package='basics',
    executable='subscriber_class',
    output='screen'
  )

  ld = LaunchDescription()
  ld.add_action(pub_cmd)
  ld.add_action(sub_cmd)

  return ld
```

br2_basics/launch/includer_launch.py

```
import os

from ament_index_python.packages import get_package_share_directory
from launch import LaunchDescription
from launch.actions import IncludeLaunchDescription
from launch.launch_description_sources import PythonLaunchDescriptionSource

def generate_launch_description():

  return LaunchDescription([
    IncludeLaunchDescription(
      PythonLaunchDescriptionSource(os.path.join(
        get_package_share_directory('basics'),
          'launch',
          'pub_sub_v2_launch.py'))
    )
  ])
```

br2_basics/launch/param_node_v1_launch.py

```
from launch import LaunchDescription
from launch_ros.actions import Node

def generate_launch_description():

  param_reader_cmd = Node(
    package='basics',
    executable='param_reader',
    parameters=[{
      'particles': 300,
      'topics': ['scan', 'image'],
      'topic_types': ['sensor_msgs/msg/LaserScan', 'sensor_msgs/msg/Image']
    }],
    output='screen'
  )

  ld = LaunchDescription()
  ld.add_action(param_reader_cmd)

  return ld
```

br2_basics/package.xml

```
<?xml version="1.0"?>
<?xml-model href="http://download.ros.org/schema/package_format3.xsd"
  schematypens="http://www.w3.org/2001/XMLSchema"?>
<package format="3">
  <name>br2_basics</name>
  <version>0.0.0</version>
  <description>Basic nodes for ROS2 introduction</description>
  <maintainer email="fmrico@gmail.com">Francisco Martin</maintainer>
  <license>Apache 2</license>

  <buildtool_depend>ament_cmake</buildtool_depend>

  <depend>rclcpp</depend>
  <depend>std_msgs</depend>

  <test_depend>ament_lint_auto</test_depend>
  <test_depend>ament_lint_common</test_depend>

  <export>
    <build_type>ament_cmake</build_type>
  </export>
```

br2_basics/config/params.yaml

```
localization_node:
  ros__parameters:
    number_particles: 300
    topics: [scan, image]
    topic_types: [sensor_msgs/msg/LaserScan, sensor_msgs/msg/Image]
```

br2_basics/src/logger_class.cpp

```
#include "rclcpp/rclcpp.hpp"

using namespace std::chrono_literals;

class LoggerNode : public rclcpp::Node
{
public:
  LoggerNode() : Node("logger_node")
  {
    counter_ = 0;
    timer_ = create_wall_timer(
      500ms, std::bind(&LoggerNode::timer_callback, this));
  }

  void timer_callback()
  {
    RCLCPP_INFO(get_logger(), "Hello %d", counter_++);
  }

private:
  rclcpp::TimerBase::SharedPtr timer_;
  int counter_;
};

int main(int argc, char * argv[]) {
  rclcpp::init(argc, argv);

  auto node = std::make_shared<LoggerNode>();

  rclcpp::spin(node);

  rclcpp::shutdown();
  return 0;
```

br2_basics/src/subscriber_class.cpp

```
#include "rclcpp/rclcpp.hpp"
#include "std_msgs/msg/int32.hpp"

using std::placeholders::_1;

class SubscriberNode : public rclcpp::Node
{
public:
  SubscriberNode() : Node("subscriber_node")
  {
    subscriber_ = create_subscription<std_msgs::msg::Int32>("int_topic", 10,
      std::bind(&SubscriberNode::callback, this, _1));
  }

  void callback(const std_msgs::msg::Int32::SharedPtr msg)
  {
    RCLCPP_INFO(get_logger(), "Hello %d", msg->data);
  }
```

br2_basics/src/subscriber_class.cpp

```
private:
  rclcpp::Subscription<std_msgs::msg::Int32>::SharedPtr subscriber_;
};

int main(int argc, char * argv[]) {
  rclcpp::init(argc, argv);

  auto node = std::make_shared<SubscriberNode>();

  rclcpp::spin(node);

  rclcpp::shutdown();
  return 0;
```

br2_basics/src/publisher.cpp

```
#include "rclcpp/rclcpp.hpp"
#include "std_msgs/msg/int32.hpp"

using namespace std::chrono_literals;

int main(int argc, char * argv[]) {
  rclcpp::init(argc, argv);

  auto node = rclcpp::Node::make_shared("publisher_node");
  auto publisher = node->create_publisher<std_msgs::msg::Int32>(
    "int_topic", 10);

  std_msgs::msg::Int32 message;
  message.data = 0;

  rclcpp::Rate loop_rate(500ms);
  while (rclcpp::ok()) {
    message.data += 1;
    publisher->publish(message);

    rclcpp::spin_some(node);
    loop_rate.sleep();
  }

  rclcpp::shutdown();
  return 0;
```

br2_basics/src/param_reader.cpp

```
#include <vector>
#include <string>

#include "rclcpp/rclcpp.hpp"

class LocalizationNode : public rclcpp::Node
{
public:
  LocalizationNode() : Node("localization_node")
  {
    declare_parameter<int>("number_particles", 200);
    declare_parameter<std::vector<std::string>>("topics", {});
    declare_parameter<std::vector<std::string>>("topic_types", {});

    get_parameter("number_particles", num_particles_);
    RCLCPP_INFO_STREAM(get_logger(), "Number of particles: " << num_particles_);

    get_parameter("topics", topics_);
    get_parameter("topic_types", topic_types_);
```

br2_basics/src/param_reader.cpp

```
    if (topics_.size() != topic_types_.size()) {
      RCLCPP_ERROR(get_logger(), "Number of topics (%zu) != number of types (%zu)",
        topics_.size(), topic_types_.size());
    } else {
      RCLCPP_INFO_STREAM(get_logger(), "Number of topics: " << topics_.size());
      for (size_t i = 0; i < topics_.size(); i++) {
        RCLCPP_INFO_STREAM(get_logger(), "\t" << topics_[i] << "\t - " << topic_types_[i]);
      }
    }
  }

private:
  int num_particles_;
  std::vector<std::string> topics_;
  std::vector<std::string> topic_types_;
};

int main(int argc, char * argv[]) {
  rclcpp::init(argc, argv);

  auto node = std::make_shared<LocalizationNode>();

  rclcpp::spin(node);

  rclcpp::shutdown();
  return 0;
}
```

br2_basics/src/executors.cpp

```
#include "rclcpp/rclcpp.hpp"

#include "std_msgs/msg/int32.hpp"

using namespace std::chrono_literals;
using std::placeholders::_1;

class PublisherNode : public rclcpp::Node
{
public:
  PublisherNode() : Node("publisher_node")
  {
    publisher_ = create_publisher<std_msgs::msg::Int32>("int_topic", 10);
    timer_ = create_wall_timer(
      500ms, std::bind(&PublisherNode::timer_callback, this));
  }

  void timer_callback()
  {
    message_.data += 1;
    publisher_->publish(message_);
  }
```

br2_basics/src/executors.cpp

```
private:
  rclcpp::Publisher<std_msgs::msg::Int32>::SharedPtr publisher_;
  rclcpp::TimerBase::SharedPtr timer_;
  std_msgs::msg::Int32 message_;
};

class SubscriberNode : public rclcpp::Node
{
public:
  SubscriberNode() : Node("subscriber_node")
  {
    subscriber_ = create_subscription<std_msgs::msg::Int32>("int_topic", 10,
      std::bind(&SubscriberNode::callback, this, _1));
  }

  void callback(const std_msgs::msg::Int32::SharedPtr msg)
  {
    RCLCPP_INFO(get_logger(), "Hello %d", msg->data);
  }

private:
  rclcpp::Subscription<std_msgs::msg::Int32>::SharedPtr subscriber_;
};

int main(int argc, char * argv[]) {
  rclcpp::init(argc, argv);

  auto node_pub = std::make_shared<PublisherNode>();
  auto node_sub = std::make_shared<SubscriberNode>();

  rclcpp::executors::SingleThreadedExecutor executor;
  // rclcpp::executors::MultiThreadedExecutor executor(
  //   rclcpp::executor::ExecutorArgs(), 8);

  executor.add_node(node_pub);
  executor.add_node(node_sub);

  executor.spin();

  rclcpp::shutdown();
  return 0;
```

br2_basics/src/publisher_class.cpp

```
#include "rclcpp/rclcpp.hpp"

#include "std_msgs/msg/int32.hpp"

using namespace std::chrono_literals;
using std::placeholders::_1;

class PublisherNode : public rclcpp::Node
{
public:
  PublisherNode() : Node("publisher_node")
  {
    publisher_ = create_publisher<std_msgs::msg::Int32>("int_topic", 10);
    timer_ = create_wall_timer(
      500ms, std::bind(&PublisherNode::timer_callback, this));
  }

  void timer_callback()
  {
    message_.data += 1;
    publisher_->publish(message_);
  }
```

br2_basics/src/publisher_class.cpp

```
private:
  rclcpp::Publisher<std_msgs::msg::Int32>::SharedPtr publisher_;
  rclcpp::TimerBase::SharedPtr timer_;
  std_msgs::msg::Int32 message_;
};

int main(int argc, char * argv[]) {
  rclcpp::init(argc, argv);

  auto node = std::make_shared<PublisherNode>();

  rclcpp::spin(node);

  rclcpp::shutdown();
  return 0;
```

br2_basics/src/logger.cpp

```
#include "rclcpp/rclcpp.hpp"

using namespace std::chrono_literals;

int main(int argc, char * argv[]) {
  rclcpp::init(argc, argv);

  auto node = rclcpp::Node::make_shared("logger_node");

  rclcpp::Rate loop_rate(500ms);
  int counter = 0;
  while (rclcpp::ok()) {
    RCLCPP_INFO(node->get_logger(), "Hello %d", counter++);

    rclcpp::spin_some(node);
    loop_rate.sleep();
  }

  rclcpp::shutdown();
  return 0;
```

A.2 PACKAGE BR2_FSM_BUMPGO_CPP

Package br2_fsm_bumpgo_cpp

```
br2_fsm_bumpgo_cpp
├── CMakeLists.txt
├── include
│   └── br2_fsm_bumpgo_cpp
│       └── BumpGoNode.hpp
├── launch
│   └── bump_and_go.launch.py
├── package.xml
└── src
    ├── br2_fsm_bumpgo_cpp
    │   └── BumpGoNode.cpp
```

br2_fsm_bumpgo_cpp/CMakeLists.txt

```
cmake_minimum_required(VERSION 3.5)
project(br2_fsm_bumpgo_cpp)

set(CMAKE_CXX_STANDARD 17)

find_package(ament_cmake REQUIRED)
find_package(rclcpp REQUIRED)
find_package(sensor_msgs REQUIRED)
find_package(geometry_msgs REQUIRED)

set(dependencies
  rclcpp
  sensor_msgs
  geometry_msgs
)

include_directories(include)

add_executable(bumpgo
  src/br2_fsm_bumpgo_cpp/BumpGoNode.cpp
  src/bumpgo_main.cpp
)
ament_target_dependencies(bumpgo ${dependencies})

install(TARGETS
  bumpgo
  ARCHIVE DESTINATION lib
  LIBRARY DESTINATION lib
  RUNTIME DESTINATION lib/${PROJECT_NAME}
)

install(DIRECTORY launch DESTINATION share/${PROJECT_NAME})

if(BUILD_TESTING)
  find_package(ament_lint_auto REQUIRED)
  ament_lint_auto_find_test_dependencies()

  set(ament_cmake_cpplint_FOUND TRUE)
  ament_lint_auto_find_test_dependencies()
endif()

ament_package()
```

br2_fsm_bumpgo_cpp/launch/bump_and_go.launch.py

```
from launch import LaunchDescription
from launch_ros.actions import Node


def generate_launch_description():

    bumpgo_cmd = Node(package='br2_fsm_bumpgo_cpp',
                      executable='bumpgo',
                      output='screen',
                      parameters=[{
                        'use_sim_time': True
                      }],
                      remappings=[
                        ('input_scan', '/scan_raw'),
                        ('output_vel', '/nav_vel')
                      ])

    ld = LaunchDescription()
    ld.add_action(bumpgo_cmd)

    return ld
```

br2_fsm_bumpgo_cpp/package.xml

```
<?xml version="1.0"?>
<?xml-model href="http://download.ros.org/schema/package_format3.xsd"
  schematypens="http://www.w3.org/2001/XMLSchema"?>
<package format="3">
  <name>br2_fsm_bumpgo_cpp</name>
  <version>0.1.0</version>
  <description>in C++</description>
  <maintainer email="fmrico@gmail.com">Francisco Martin</maintainer>
  <license>Apache 2.0</license>

  <buildtool_depend>ament_cmake</buildtool_depend>

  <depend>rclcpp</depend>
  <depend>geometry_msgs</depend>
  <depend>sensor_msgs</depend>

  <test_depend>ament_lint_auto</test_depend>
  <test_depend>ament_lint_common</test_depend>

  <export>
    <build_type>ament_cmake</build_type>
  </export>
</package>
```

br2_fsm_bumpgo_cpp/include/br2_fsm_bumpgo_cpp/BumpGoNode.hpp

```
#ifndef BR2_BT_BUMPGO__BUMPGONODE_HPP_
#define BR2_BT_BUMPGO__BUMPGONODE_HPP_

#include "sensor_msgs/msg/laser_scan.hpp"
#include "geometry_msgs/msg/twist.hpp"

#include "rclcpp/rclcpp.hpp"

namespace br2_fsm_bumpgo_cpp
{

using namespace std::chrono_literals;

class BumpGoNode : public rclcpp::Node
{
public:
  BumpGoNode();

private:
  void scan_callback(sensor_msgs::msg::LaserScan::UniquePtr msg);
  void control_cycle();

  static const int FORWARD = 0;
  static const int BACK = 1;
  static const int TURN = 2;
  static const int STOP = 3;
  int state_;
  rclcpp::Time state_ts_;

  void go_state(int new_state);
  bool check_forward_2_back();
  bool check_forward_2_stop();
  bool check_back_2_turn();
  bool check_turn_2_forward();
  bool check_stop_2_forward();

  const rclcpp::Duration TURNING_TIME {2s};
  const rclcpp::Duration BACKING_TIME {2s};
  const rclcpp::Duration SCAN_TIMEOUT {1s};
```

br2_fsm_bumpgo_cpp/include/br2_fsm_bumpgo_cpp/BumpGoNode.hpp

```
  static constexpr float SPEED_LINEAR = 0.3f;
  static constexpr float SPEED_ANGULAR = 0.3f;
  static constexpr float OBSTACLE_DISTANCE = 1.0f;

  rclcpp::Subscription<sensor_msgs::msg::LaserScan>::SharedPtr scan_sub_;
  rclcpp::Publisher<geometry_msgs::msg::Twist>::SharedPtr vel_pub_;
  rclcpp::TimerBase::SharedPtr timer_;

  sensor_msgs::msg::LaserScan::UniquePtr last_scan_;
};

} // namespace br2_fsm_bumpgo_cpp

#endif // BR2_BT_BUMPGO__BUMPGONODE_HPP_
```

br2_fsm_bumpgo_cpp/src/bumpgo_main.cpp

```
#include <memory>

#include "br2_fsm_bumpgo_cpp/BumpGoNode.hpp"
#include "rclcpp/rclcpp.hpp"

int main(int argc, char * argv[])
{
  rclcpp::init(argc, argv);

  auto bumpgo_node = std::make_shared<br2_fsm_bumpgo_cpp::BumpGoNode>();
  rclcpp::spin(bumpgo_node);

  rclcpp::shutdown();

  return 0;
}
```

br2_fsm_bumpgo_cpp/src/br2_fsm_bumpgo_cpp/BumpGoNode.cpp

```
#include <utility>
#include "br2_fsm_bumpgo_cpp/BumpGoNode.hpp"

#include "sensor_msgs/msg/laser_scan.hpp"
#include "geometry_msgs/msg/twist.hpp"

#include "rclcpp/rclcpp.hpp"

namespace br2_fsm_bumpgo_cpp
{

using namespace std::chrono_literals;
using std::placeholders::_1;

BumpGoNode::BumpGoNode()
: Node("bump_go"),
  state_(FORWARD)
{
  scan_sub_ = create_subscription<sensor_msgs::msg::LaserScan>(
    "input_scan", rclcpp::SensorDataQoS(),
    std::bind(&BumpGoNode::scan_callback, this, _1));

  vel_pub_ = create_publisher<geometry_msgs::msg::Twist>("output_vel", 10);
  timer_ = create_wall_timer(50ms, std::bind(&BumpGoNode::control_cycle, this));

  state_ts_ = now();
}

void
BumpGoNode::scan_callback(sensor_msgs::msg::LaserScan::UniquePtr msg)
{
  last_scan_ = std::move(msg);
}
```

br2_fsm_bumpgo_cpp/src/br2_fsm_bumpgo_cpp/BumpGoNode.cpp

```
void
BumpGoNode::control_cycle()
{
  // Do nothing until the first sensor read
  if (last_scan_ == nullptr)
    return;

  geometry_msgs::msg::Twist out_vel;

  switch (state_) {
    case FORWARD:
      out_vel.linear.x = SPEED_LINEAR;

      if (check_forward_2_stop())
        go_state(STOP);
      if (check_forward_2_back())
        go_state(BACK);

      break;
    case BACK:
      out_vel.linear.x = -SPEED_LINEAR;

      if (check_back_2_turn())
        go_state(TURN);

      break;
    case TURN:
      out_vel.angular.z = SPEED_ANGULAR;

      if (check_turn_2_forward())
        go_state(FORWARD);

      break;
    case STOP:
      if (check_stop_2_forward())
        go_state(FORWARD);
      break;
  }

  vel_pub_->publish(out_vel);
}

void
BumpGoNode::go_state(int new_state)
{
  state_ = new_state;
  state_ts_ = now();
}

bool
BumpGoNode::check_forward_2_back()
{
  // going forward when deteting an obstacle
  // at 0.5 meters with the front laser read
  size_t pos = last_scan_->ranges.size() / 2;
  return last_scan_->ranges[pos] < OBSTACLE_DISTANCE;
}

bool
BumpGoNode::check_forward_2_stop()
{
  // Stop if no sensor readings for 1 second
  auto elapsed = now() - rclcpp::Time(last_scan_->header.stamp);
  return elapsed > SCAN_TIMEOUT;
}

bool
BumpGoNode::check_stop_2_forward()
{
  // Going forward if sensor readings are available
  // again
  auto elapsed = now() - rclcpp::Time(last_scan_->header.stamp);
  return elapsed < SCAN_TIMEOUT;
}
```

br2_fsm_bumpgo_cpp/src/br2_fsm_bumpgo_cpp/BumpGoNode.cpp

```
bool
BumpGoNode::check_back_2_turn()
{
  // Going back for 2 seconds
  return (now() - state_ts_) > BACKING_TIME;
}

bool
BumpGoNode::check_turn_2_forward()
{
  // Turning for 2 seconds
  return (now() - state_ts_) > TURNING_TIME;
}
```

A.3 PACKAGE BR2_FSM_BUMPGO_PY

Package br2_fsm_bumpgo_py

```
br2_fsm_bumpgo_py
├── br2_fsm_bumpgo_py
│   ├── bump_go_main.py
│   └── __init__.py
├── launch
│   └── bump_and_go.launch.py
├── package.xml
├── resource
│   └── br2_fsm_bumpgo_py
├── setup.cfg
├── setup.py
└── test
    ├── test_copyright.py
    ├── test_flake8.py
    └── test_pep257.py
```

br2_fsm_bumpgo_py/launch/bump_and_go.launch.py

```
from launch import LaunchDescription
from launch_ros.actions import Node


def generate_launch_description():

    kobuki_cmd = Node(package='br2_fsm_bumpgo_py',
                      executable='bump_go_main',
                      output='screen',
                      parameters=[{
                        'use_sim_time': True
                      }],
                      remappings=[
                        ('input_scan', '/scan_raw'),
                        ('output_vel', '/nav_vel')
                      ])

    ld = LaunchDescription()
    ld.add_action(kobuki_cmd)

    return ld
```

br2_fsm_bumpgo_py/package.xml

```
<?xml version="1.0"?>
<?xml-model href="http://download.ros.org/schema/package_format3.xsd"
  schematypens="http://www.w3.org/2001/XMLSchema"?>
<package format="3">
  <name>br2_fsm_bumpgo_py</name>
  <version>0.0.0</version>
  <description>Bump and Go behavior based on Finite State Machines in Python</description>
  <maintainer email="fmrico@gmail.com">Francisco Martin</maintainer>
  <license>Apache 2.0</license>

  <depend>rclcpy</depend>
  <depend>sensor_msgs</depend>
  <depend>geometry_msgs</depend>

  <test_depend>ament_copyright</test_depend>
  <test_depend>ament_flake8</test_depend>
  <test_depend>ament_pep257</test_depend>
  <test_depend>python3-pytest</test_depend>

  <export>
    <build_type>ament_python</build_type>
  </export>
</package>
```

br2_fsm_bumpgo_py/setup.py

```
import os
from glob import glob

from setuptools import setup

package_name = 'br2_fsm_bumpgo_py'

setup(
    name=package_name,
    version='0.0.0',
    packages=[package_name],
    data_files=[
        ('share/ament_index/resource_index/packages',
            ['resource/' + package_name]),
        ('share/' + package_name, ['package.xml']),
        (os.path.join('share', package_name, 'launch'), glob('launch/*.launch.py'))
    ],
    install_requires=['setuptools'],
    zip_safe=True,
    maintainer='fmrico',
    maintainer_email='fmrico@gmail.com',
    description='Bump and Go behavior based on Finite State Machines in Python',
    license='Apache 2.0',
    tests_require=['pytest'],
    entry_points={
        'console_scripts': [
          'bump_go_main = br2_fsm_bumpgo_py.bump_go_main:main'
        ],
    },
)
```

br2_fsm_bumpgo_py/bump_go_py.py

```
import rclpy

from rclpy.duration import Duration
from rclpy.node import Node
from rclpy.qos import qos_profile_sensor_data
from rclpy.time import Time

from geometry_msgs.msg import Twist
from sensor_msgs.msg import LaserScan


class BumpGoNode(Node):
    def __init__(self):
        super().__init__('bump_go')

        self.FORWARD = 0
        self.BACK = 1
        self.TURN = 2
        self.STOP = 3
        self.state = self.FORWARD
        self.state_ts = self.get_clock().now()

        self.TURNING_TIME = 2.0
        self.BACKING_TIME = 2.0
        self.SCAN_TIMEOUT = 1.0

        self.SPEED_LINEAR = 0.3
        self.SPEED_ANGULAR = 0.3
        self.OBSTACLE_DISTANCE = 1.0

        self.last_scan = None

        self.scan_sub = self.create_subscription(
            LaserScan,
            'input_scan',
            self.scan_callback,
            qos_profile_sensor_data)

        self.vel_pub = self.create_publisher(Twist, 'output_vel', 10)
        self.timer = self.create_timer(0.05, self.control_cycle)

    def scan_callback(self, msg):
        self.last_scan = msg

    def control_cycle(self):
        if self.last_scan is None:
            return

        out_vel = Twist()

        if self.state == self.FORWARD:
          out_vel.linear.x = self.SPEED_LINEAR

          if self.check_forward_2_stop():
            self.go_state(self.STOP)
          if self.check_forward_2_back():
            self.go_state(self.BACK)

        elif self.state == self.BACK:
          out_vel.linear.x = -self.SPEED_LINEAR

          if self.check_back_2_turn():
            self.go_state(self.TURN)

        elif self.state == self.TURN:
          out_vel.angular.z = self.SPEED_ANGULAR

          if self.check_turn_2_forward():
            self.go_state(self.FORWARD)

        elif self.state == self.STOP:
          if self.check_stop_2_forward():
            self.go_state(self.FORWARD)

        self.vel_pub.publish(out_vel)

    def go_state(self, new_state):
        self.state = new_state
        self.state_ts = self.get_clock().now()
```

br2_fsm_bumpgo_py/setup.py

```
    def check_forward_2_back(self):
        pos = round(len(self.last_scan.ranges) / 2)
        return self.last_scan.ranges[pos] < self.OBSTACLE_DISTANCE

    def check_forward_2_stop(self):
        elapsed = self.get_clock().now() - Time.from_msg(self.last_scan.header.stamp)
        return elapsed > Duration(seconds=self.SCAN_TIMEOUT)

    def check_stop_2_forward(self):
        elapsed = self.get_clock().now() - Time.from_msg(self.last_scan.header.stamp)
        return elapsed < Duration(seconds=self.SCAN_TIMEOUT)

    def check_back_2_turn(self):
        elapsed = self.get_clock().now() - self.state_ts
        return elapsed > Duration(seconds=self.BACKING_TIME)

    def check_turn_2_forward(self):
        elapsed = self.get_clock().now() - self.state_ts
        return elapsed > Duration(seconds=self.TURNING_TIME)

def main(args=None):
    rclpy.init(args=args)

    bump_go_node = BumpGoNode()

    rclpy.spin(bump_go_node)

    bump_go_node.destroy_node()
    rclpy.shutdown()

if __name__ == '__main__':
    main()
```

A.4 PACKAGE BR2_TF2_DETECTOR

Package br2_tf2_detector

```
br2_tf2_detector
├── CMakeLists.txt
├── include
│   └── br2_tf2_detector
│       ├── ObstacleDetectorImprovedNode.hpp
│       ├── ObstacleDetectorNode.hpp
│       └── ObstacleMonitorNode.hpp
├── launch
│   ├── detector_basic.launch.py
│   └── detector_improved.launch.py
├── package.xml
└── src
    ├── br2_tf2_detector
    │   ├── ObstacleDetectorImprovedNode.cpp
    │   ├── ObstacleDetectorNode.cpp
    │   └── ObstacleMonitorNode.cpp
    ├── detector_improved_main.cpp
```

br2_tf2_detector/CMakeLists.txt

```
cmake_minimum_required(VERSION 3.5)
project(br2_tf2_detector)

set(CMAKE_CXX_STANDARD 17)

# find dependencies
find_package(ament_cmake REQUIRED)
find_package(rclcpp REQUIRED)
find_package(tf2_ros REQUIRED)
find_package(geometry_msgs REQUIRED)
find_package(sensor_msgs REQUIRED)
find_package(visualization_msgs REQUIRED)

set(dependencies
    rclcpp
    tf2_ros
    geometry_msgs
    sensor_msgs
    visualization_msgs
)

include_directories(include)

add_library(${PROJECT_NAME} SHARED
  src/br2_tf2_detector/ObstacleDetectorNode.cpp
  src/br2_tf2_detector/ObstacleMonitorNode.cpp
  src/br2_tf2_detector/ObstacleDetectorImprovedNode.cpp
)
ament_target_dependencies(${PROJECT_NAME} ${dependencies})

add_executable(detector src/detector_main.cpp)
ament_target_dependencies(detector ${dependencies})
target_link_libraries(detector ${PROJECT_NAME})

add_executable(detector_improved src/detector_improved_main.cpp)
ament_target_dependencies(detector_improved ${dependencies})
target_link_libraries(detector_improved ${PROJECT_NAME})

install(TARGETS
  ${PROJECT_NAME}
  detector
  detector_improved
  ARCHIVE DESTINATION lib
  LIBRARY DESTINATION lib
  RUNTIME DESTINATION lib/${PROJECT_NAME}
)

install(DIRECTORY launch DESTINATION share/${PROJECT_NAME})

if(BUILD_TESTING)
  find_package(ament_lint_auto REQUIRED)
  ament_lint_auto_find_test_dependencies()
endif()

ament_package()
```

br2_tf2_detector/launch/detector_improved.launch.py

```
from launch import LaunchDescription
from launch_ros.actions import Node


def generate_launch_description():

    detector_cmd = Node(package='br2_tf2_detector',
                        executable='detector_improved',
                        output='screen',
                        parameters=[{
                          'use_sim_time': True
                        }],
                        remappings=[
                          ('input_scan', '/scan_raw')
                        ])

    ld = LaunchDescription()
    ld.add_action(detector_cmd)

    return ld
```

br2_tf2_detector/launch/detector_basic.launch.py

```
from launch import LaunchDescription
from launch_ros.actions import Node


def generate_launch_description():

    detector_cmd = Node(package='br2_tf2_detector',
                        executable='detector',
                        output='screen',
                        parameters=[{
                          'use_sim_time': True
                        }],
                        remappings=[
                          ('input_scan', '/scan_raw')
                        ])

    ld = LaunchDescription()
    ld.add_action(detector_cmd)

    return ld
```

br2_tf2_detector/package.xml

```
<?xml version="1.0"?>
<?xml-model href="http://download.ros.org/schema/package_format3.xsd"
  schematypens="http://www.w3.org/2001/XMLSchema"?>
<package format="3">
  <name>br2_tf2_detector</name>
  <version>0.0.0</version>
  <description>TF detector package</description>
  <maintainer email="fmrico@gmail.com">Francisco Martín</maintainer>
  <license>Apache 2.0</license>

  <buildtool_depend>ament_cmake</buildtool_depend>

  <depend>rclcpp</depend>
  <depend>tf2_ros</depend>
  <depend>geometry_msgs</depend>
  <depend>sensor_msgs</depend>

  <test_depend>ament_lint_auto</test_depend>
  <test_depend>ament_lint_common</test_depend>

  <export>
    <build_type>ament_cmake</build_type>
  </export>
</package>
```

br2_tf2_detector/include/br2_tf2_detector/ObstacleDetectorImprovedNode.hpp

```
#ifndef BR2_TF2_DETECTOR__OBSTACLEDETECTORIMPROVEDNODE_HPP_
#define BR2_TF2_DETECTOR__OBSTACLEDETECTORIMPROVEDNODE_HPP_

#include <tf2_ros/static_transform_broadcaster.h>
#include <tf2_ros/buffer.h>
#include <tf2_ros/transform_listener.h>

#include <memory>

#include "sensor_msgs/msg/laser_scan.hpp"

#include "rclcpp/rclcpp.hpp"

namespace br2_tf2_detector
{

class ObstacleDetectorImprovedNode : public rclcpp::Node
{
public:
  ObstacleDetectorImprovedNode();

private:
  void scan_callback(sensor_msgs::msg::LaserScan::UniquePtr msg);

  rclcpp::Subscription<sensor_msgs::msg::LaserScan>::SharedPtr scan_sub_;
  std::shared_ptr<tf2_ros::StaticTransformBroadcaster> tf_broadcaster_;

  tf2::BufferCore tf_buffer_;
  tf2_ros::TransformListener tf_listener_;
};

}  // namespace br2_tf2_detector
#endif  // BR2_TF2_DETECTOR__OBSTACLEDETECTORIMPROVEDNODE_HPP_
```

br2_tf2_detector/include/br2_tf2_detector/ObstacleMonitorNode.hpp

```
#ifndef BR2_TF2_DETECTOR__OBSTACLEMONITORNODE_HPP_
#define BR2_TF2_DETECTOR__OBSTACLEMONITORNODE_HPP_

#include <tf2_ros/buffer.h>
#include <tf2_ros/transform_listener.h>

#include <memory>


#include "sensor_msgs/msg/laser_scan.hpp"
#include "visualization_msgs/msg/marker.hpp"

#include "rclcpp/rclcpp.hpp"

namespace br2_tf2_detector
{

class ObstacleMonitorNode : public rclcpp::Node
{
public:
  ObstacleMonitorNode();

private:
  void control_cycle();
  rclcpp::TimerBase::SharedPtr timer_;

  tf2::BufferCore tf_buffer_;
  tf2_ros::TransformListener tf_listener_;

  rclcpp::Publisher<visualization_msgs::msg::Marker>::SharedPtr marker_pub_;
};

}  // namespace br2_tf2_detector
#endif  // BR2_TF2_DETECTOR__OBSTACLEMONITORNODE_HPP_
```

br2_tf2_detector/include/br2_tf2_detector/ObstacleDetectorNode.hpp

```
#ifndef BR2_TF2_DETECTOR__OBSTACLEDETECTORNODE_HPP_
#define BR2_TF2_DETECTOR__OBSTACLEDETECTORNODE_HPP_

#include <tf2_ros/static_transform_broadcaster.h>

#include <memory>

#include "sensor_msgs/msg/laser_scan.hpp"

#include "rclcpp/rclcpp.hpp"

namespace br2_tf2_detector
{

class ObstacleDetectorNode : public rclcpp::Node
{
public:
  ObstacleDetectorNode();

private:
  void scan_callback(sensor_msgs::msg::LaserScan::UniquePtr msg);

  rclcpp::Subscription<sensor_msgs::msg::LaserScan>::SharedPtr scan_sub_;
  std::shared_ptr<tf2_ros::StaticTransformBroadcaster> tf_broadcaster_;
};

}  // namespace br2_tf2_detector
#endif  // BR2_TF2_DETECTOR__OBSTACLEDETECTORNODE_HPP_
```

br2_tf2_detector/src/detector_main.cpp

```
#include <memory>

#include "br2_tf2_detector/ObstacleDetectorNode.hpp"
#include "br2_tf2_detector/ObstacleMonitorNode.hpp"

#include "rclcpp/rclcpp.hpp"

int main(int argc, char * argv[])
{
  rclcpp::init(argc, argv);

  auto obstacle_detector = std::make_shared<br2_tf2_detector::ObstacleDetectorNode>();
  auto obstacle_monitor = std::make_shared<br2_tf2_detector::ObstacleMonitorNode>();

  rclcpp::executors::SingleThreadedExecutor executor;
  executor.add_node(obstacle_detector->get_node_base_interface());
  executor.add_node(obstacle_monitor->get_node_base_interface());

  executor.spin();

  rclcpp::shutdown();
  return 0;
}
```

br2_tf2_detector/src/detector_improved_main.cpp

```
#include <memory>

#include "br2_tf2_detector/ObstacleDetectorImprovedNode.hpp"
#include "br2_tf2_detector/ObstacleMonitorNode.hpp"

#include "rclcpp/rclcpp.hpp"

int main(int argc, char * argv[])
{
  rclcpp::init(argc, argv);

  auto obstacle_detector = std::make_shared<br2_tf2_detector::
    ObstacleDetectorImprovedNode>();
  auto obstacle_monitor = std::make_shared<br2_tf2_detector::ObstacleMonitorNode>();

  rclcpp::executors::SingleThreadedExecutor executor;
  executor.add_node(obstacle_detector->get_node_base_interface());
  executor.add_node(obstacle_monitor->get_node_base_interface());

  executor.spin();

  rclcpp::shutdown();
  return 0;
}
```

br2_tf2_detector/src/br2_tf2_detector/ObstacleMonitorNode.cpp

```
#include <tf2/transform_datatypes.h>
#include <tf2/LinearMath/Quaternion.h>
#include <tf2_geometry_msgs/tf2_geometry_msgs.h>

#include <memory>

#include "br2_tf2_detector/ObstacleMonitorNode.hpp"

#include "geometry_msgs/msg/transform_stamped.hpp"

#include "rclcpp/rclcpp.hpp"

namespace br2_tf2_detector
{

using namespace std::chrono_literals;

ObstacleMonitorNode::ObstacleMonitorNode()
: Node("obstacle_monitor"),
  tf_buffer_(),
  tf_listener_(tf_buffer_)
{
  marker_pub_ = create_publisher<visualization_msgs::msg::Marker>("obstacle_marker", 1);

  timer_ = create_wall_timer(
    500ms, std::bind(&ObstacleMonitorNode::control_cycle, this));
}

void
ObstacleMonitorNode::control_cycle()
{
  geometry_msgs::msg::TransformStamped robot2obstacle;

  try {
    robot2obstacle = tf_buffer_.lookupTransform(
      "base_footprint", "detected_obstacle", tf2::TimePointZero);
  } catch (tf2::TransformException & ex) {
    RCLCPP_WARN(get_logger(), "Obstacle transform not found: %s", ex.what());
    return;
  }

  double x = robot2obstacle.transform.translation.x;
  double y = robot2obstacle.transform.translation.y;
  double z = robot2obstacle.transform.translation.z;
  double theta = atan2(y, x);

  RCLCPP_INFO(
    get_logger(), "Obstacle detected at (%lf m, %lf m, , %lf m) = %lf rads",
    x, y, z, theta);

  visualization_msgs::msg::Marker obstacle_arrow;
  obstacle_arrow.header.frame_id = "base_footprint";
  obstacle_arrow.header.stamp = now();
  obstacle_arrow.type = visualization_msgs::msg::Marker::ARROW;
  obstacle_arrow.action = visualization_msgs::msg::Marker::ADD;
  obstacle_arrow.lifetime = rclcpp::Duration(1s);

  geometry_msgs::msg::Point start;
  start.x = 0.0;
  start.y = 0.0;
  start.z = 0.0;
  geometry_msgs::msg::Point end;
  end.x = x;
  end.y = y;
  end.z = z;
  obstacle_arrow.points = {start, end};

  obstacle_arrow.color.r = 1.0;
  obstacle_arrow.color.g = 0.0;
  obstacle_arrow.color.b = 0.0;
  obstacle_arrow.color.a = 1.0;

  obstacle_arrow.scale.x = 0.02;
  obstacle_arrow.scale.y = 0.1;
  obstacle_arrow.scale.z = 0.1;
```

br2_tf2_detector/src/br2_tf2_detector/ObstacleMonitorNode.cpp

```
  marker_pub_->publish(obstacle_arrow);
}

}  // namespace br2_tf2_detector
```

br2_tf2_detector/src/br2_tf2_detector/ObstacleDetectorNode.cpp

```
#include <memory>

#include "br2_tf2_detector/ObstacleDetectorNode.hpp"

#include "sensor_msgs/msg/laser_scan.hpp"
#include "geometry_msgs/msg/transform_stamped.hpp"

#include "rclcpp/rclcpp.hpp"

namespace br2_tf2_detector
{

using std::placeholders::_1;

ObstacleDetectorNode::ObstacleDetectorNode()
: Node("obstacle_detector")
{
  scan_sub_ = create_subscription<sensor_msgs::msg::LaserScan>(
    "input_scan", rclcpp::SensorDataQoS(),
    std::bind(&ObstacleDetectorNode::scan_callback, this, _1));

  tf_broadcaster_ = std::make_shared<tf2_ros::StaticTransformBroadcaster>(*this);
}

void
ObstacleDetectorNode::scan_callback(sensor_msgs::msg::LaserScan::UniquePtr msg)
{
  double dist = msg->ranges[msg->ranges.size() / 2];

  if (!std::isinf(dist)) {
    geometry_msgs::msg::TransformStamped detection_tf;

    detection_tf.header = msg->header;
    detection_tf.child_frame_id = "detected_obstacle";
    detection_tf.transform.translation.x = msg->ranges[msg->ranges.size() / 2];

    tf_broadcaster_->sendTransform(detection_tf);
  }
}

}  // namespace br2_tf2_detector
```

br2_tf2_detector/src/br2_tf2_detector/ObstacleDetectorImprovedNode.cpp

```
#include <tf2/transform_datatypes.h>
#include <tf2/LinearMath/Quaternion.h>
#include <tf2_geometry_msgs/tf2_geometry_msgs.h>

#include <memory>

#include "br2_tf2_detector/ObstacleDetectorImprovedNode.hpp"

#include "sensor_msgs/msg/laser_scan.hpp"
#include "geometry_msgs/msg/transform_stamped.hpp"

#include "rclcpp/rclcpp.hpp"

namespace br2_tf2_detector
{

using std::placeholders::_1;
using namespace std::chrono_literals;
```

br2_tf2_detector/src/br2_tf2_detector/ObstacleDetectorImprovedNode.cpp

```
ObstacleDetectorImprovedNode::ObstacleDetectorImprovedNode()
: Node("obstacle_detector_improved"),
  tf_buffer_(),
  tf_listener_(tf_buffer_)
{
  scan_sub_ = create_subscription<sensor_msgs::msg::LaserScan>(
    "input_scan", rclcpp::SensorDataQoS(),
    std::bind(&ObstacleDetectorImprovedNode::scan_callback, this, _1));

  tf_broadcaster_ = std::make_shared<tf2_ros::StaticTransformBroadcaster>(*this);
}

void
ObstacleDetectorImprovedNode::scan_callback(sensor_msgs::msg::LaserScan::UniquePtr msg)
{
  double dist = msg->ranges[msg->ranges.size() / 2];

  if (!std::isinf(dist)) {
    tf2::Transform laser2object;
    laser2object.setOrigin(tf2::Vector3(dist, 0.0, 0.0));
    laser2object.setRotation(tf2::Quaternion(0.0, 0.0, 0.0, 1.0));

    geometry_msgs::msg::TransformStamped odom2laser_msg;
    tf2::Stamped<tf2::Transform> odom2laser;
    try {
      odom2laser_msg = tf_buffer_.lookupTransform(
        "odom", "base_laser_link",
        tf2::timeFromSec(rclcpp::Time(msg->header.stamp).seconds()));
      tf2::fromMsg(odom2laser_msg, odom2laser);
    } catch (tf2::TransformException & ex) {
      RCLCPP_WARN(get_logger(), "Obstacle transform not found: %s", ex.what());
      return;
    }

    tf2::Transform odom2object = odom2laser * laser2object;

    geometry_msgs::msg::TransformStamped odom2object_msg;
    odom2object_msg.transform = tf2::toMsg(odom2object);

    odom2object_msg.header.stamp = msg->header.stamp;
    odom2object_msg.header.frame_id = "odom";
    odom2object_msg.child_frame_id = "detected_obstacle";

    tf_broadcaster_->sendTransform(odom2object_msg);
  }
}

}  // namespace br2_tf2_detector
```

A.5 PACKAGE BR2_VFF_AVOIDANCE

Package br2_vff_avoidance

```
br2_vff_avoidance
├── CMakeLists.txt
├── include
│   └── br2_vff_avoidance
│       └── AvoidanceNode.hpp
├── launch
│   └── avoidance_vff.launch.py
├── package.xml
├── src
│   ├── avoidance_vff_main.cpp
│   └── br2_vff_avoidance
│       └── AvoidanceNode.cpp
└── tests
    ├── CMakeLists.txt
    └── vff_test.cpp
```

br2_vff_avoidance/CMakeLists.txt

```
cmake_minimum_required(VERSION 3.5)
project(br2_vff_avoidance)

set(CMAKE_CXX_STANDARD 17)
set(CMAKE_BUILD_TYPE Debug)

find_package(ament_cmake REQUIRED)
find_package(rclcpp REQUIRED)
find_package(sensor_msgs REQUIRED)
find_package(geometry_msgs REQUIRED)
find_package(visualization_msgs REQUIRED)

set(dependencies
    rclcpp
    sensor_msgs
    geometry_msgs
    visualization_msgs
)

include_directories(include)

add_library(${PROJECT_NAME} SHARED src/br2_vff_avoidance/AvoidanceNode.cpp)
ament_target_dependencies(${PROJECT_NAME} ${dependencies})

add_executable(avoidance_vff src/avoidance_vff_main.cpp)
ament_target_dependencies(avoidance_vff ${dependencies})
target_link_libraries(avoidance_vff ${PROJECT_NAME})

install(TARGETS
  ${PROJECT_NAME}
  avoidance_vff
  ARCHIVE DESTINATION lib
  LIBRARY DESTINATION lib
  RUNTIME DESTINATION lib/${PROJECT_NAME}
)

install(DIRECTORY launch DESTINATION share/${PROJECT_NAME})

if(BUILD_TESTING)
  find_package(ament_lint_auto REQUIRED)
  ament_lint_auto_find_test_dependencies()

  set(ament_cmake_cpplint_FOUND TRUE)
  ament_lint_auto_find_test_dependencies()

  find_package(ament_cmake_gtest REQUIRED)
  add_subdirectory(tests)
endif()

ament_export_dependencies(${dependencies})
ament_package()
```

br2_vff_avoidance/launch/avoidance_vff.launch.py

```
from launch import LaunchDescription
from launch_ros.actions import Node


def generate_launch_description():

    vff_avoidance_cmd = Node(
      package='br2_vff_avoidance',
      executable='avoidance_vff',
      parameters=[{
        'use_sim_time': True
      }],
      remappings=[
        ('input_scan', '/scan_raw'),
        ('output_vel', '/nav_vel')
      ],
      output='screen'
    )

    ld = LaunchDescription()
    ld.add_action(vff_avoidance_cmd)

    return ld
```

br2_vff_avoidance/package.xml

```
<?xml version="1.0"?>
<?xml-model href="http://download.ros.org/schema/package_format3.xsd"
  schematypens="http://www.w3.org/2001/XMLSchema"?>
<package format="3">
  <name>br2_vff_avoidance</name>
  <version>0.1.0</version>
  <description>VFF Avoidance package</description>
  <maintainer email="fmrico@gmail.com">Francisco Martín</maintainer>
  <license>Apache 2.0</license>

  <buildtool_depend>ament_cmake</buildtool_depend>

  <depend>rclcpp</depend>
  <depend>geometry_msgs</depend>
  <depend>sensor_msgs</depend>
  <depend>visualization_msgs</depend>

  <test_depend>ament_lint_auto</test_depend>
  <test_depend>ament_lint_common</test_depend>
  <test_depend>ament_cmake_gtest</test_depend>

  <export>
    <build_type>ament_cmake</build_type>
  </export>
</package>
```

br2_vff_avoidance/include/br2_vff_avoidance/AvoidanceNode.hpp

```
#ifndef BR2_VFF_AVOIDANCE__AVOIDANCENODE_HPP_
#define BR2_VFF_AVOIDANCE__AVOIDANCENODE_HPP_

#include <memory>
#include <vector>

#include "geometry_msgs/msg/twist.hpp"
#include "sensor_msgs/msg/laser_scan.hpp"
#include "visualization_msgs/msg/marker_array.hpp"

#include "rclcpp/rclcpp.hpp"

namespace br2_vff_avoidance
{
```

br2_vff_avoidance/include/br2_vff_avoidance/AvoidanceNode.hpp

```
struct VFFVectors
{
  std::vector<float> attractive;
  std::vector<float> repulsive;
  std::vector<float> result;
};

typedef enum {RED, GREEN, BLUE, NUM_COLORS} VFFColor;

class AvoidanceNode : public rclcpp::Node
{
public:
  AvoidanceNode();

  void scan_callback(sensor_msgs::msg::LaserScan::UniquePtr msg);
  void control_cycle();

protected:
  VFFVectors get_vff(const sensor_msgs::msg::LaserScan & scan);

  visualization_msgs::msg::MarkerArray get_debug_vff(const VFFVectors & vff_vectors);
  visualization_msgs::msg::Marker make_marker(
    const std::vector<float> & vector, VFFColor vff_color);

private:
  rclcpp::Publisher<geometry_msgs::msg::Twist>::SharedPtr vel_pub_;
  rclcpp::Publisher<visualization_msgs::msg::MarkerArray>::SharedPtr vff_debug_pub_;
  rclcpp::Subscription<sensor_msgs::msg::LaserScan>::SharedPtr scan_sub_;
  rclcpp::TimerBase::SharedPtr timer_;

  sensor_msgs::msg::LaserScan::UniquePtr last_scan_;
};

}  // namespace br2_vff_avoidance

#endif  // BR2_VFF_AVOIDANCE__AVOIDANCENODE_HPP_
```

br2_vff_avoidance/src/br2_vff_avoidance/AvoidanceNode.cpp

```
#include <memory>
#include <utility>
#include <algorithm>
#include <vector>

#include "geometry_msgs/msg/twist.hpp"
#include "sensor_msgs/msg/laser_scan.hpp"
#include "visualization_msgs/msg/marker_array.hpp"

#include "br2_vff_avoidance/AvoidanceNode.hpp"

#include "rclcpp/rclcpp.hpp"

using std::placeholders::_1;
using namespace std::chrono_literals;

namespace br2_vff_avoidance
{

AvoidanceNode::AvoidanceNode()
: Node("avoidance_vff")
{
  vel_pub_ = create_publisher<geometry_msgs::msg::Twist>("output_vel", 100);
  vff_debug_pub_ = create_publisher<visualization_msgs::msg::MarkerArray>("vff_debug",
    100);

  scan_sub_ = create_subscription<sensor_msgs::msg::LaserScan>(
    "input_scan", rclcpp::SensorDataQoS(), std::bind(&AvoidanceNode::scan_callback,
      this, _1));

  timer_ = create_wall_timer(50ms, std::bind(&AvoidanceNode::control_cycle, this));
}
```

br2_vff_avoidance/src/br2_vff_avoidance/AvoidanceNode.cpp

```
void
AvoidanceNode::scan_callback(sensor_msgs::msg::LaserScan::UniquePtr msg)
{
  last_scan_ = std::move(msg);
}

void
AvoidanceNode::control_cycle()
{
  // Skip cycle if no valid recent scan available
  if (last_scan_ == nullptr || (now() - last_scan_->header.stamp) > 1s) {
    return;
  }

  // Get VFF vectors
  const VFFVectors & vff = get_vff(*last_scan_);

  // Use result vector to calculate output speed
  const auto & v = vff.result;
  double angle = atan2(v[1], v[0]);
  double module = sqrt(v[0] * v[0] + v[1] * v[1]);

  // Create ouput message, controlling speed limits
  geometry_msgs::msg::Twist vel;
  vel.linear.x = std::clamp(module, 0.0, 0.3);  // truncate linear vel to [0.0, 0.3] m/s
  vel.angular.z = std::clamp(angle, -0.5, 0.5);  // truncate rot vel to [-0.5, 0.5] rad/s

  vel_pub_->publish(vel);

  // Produce debug information, if any interested
  if (vff_debug_pub_->get_subscription_count() > 0) {
    vff_debug_pub_->publish(get_debug_vff(vff));
  }
}

VFFVectors
AvoidanceNode::get_vff(const sensor_msgs::msg::LaserScan & scan)
{
  // This is the obstacle radious in which an obstacle affects the robot
  const float OBSTACLE_DISTANCE = 1.0;

  // Init vectors
  VFFVectors vff_vector;
  vff_vector.attractive = {OBSTACLE_DISTANCE, 0.0};  // Robot wants to go forward
  vff_vector.repulsive = {0.0, 0.0};
  vff_vector.result = {0.0, 0.0};

  // Get the index of nearest obstacle
  int min_idx = std::min_element(scan.ranges.begin(), scan.ranges.end()) -
    scan.ranges.begin();

  // Get the distance to nearest obstacle
  float distance_min = scan.ranges[min_idx];

  // If the obstacle is in the area that affects the robot, calculate repulsive vector
  if (distance_min < OBSTACLE_DISTANCE) {
    float angle = scan.angle_min + scan.angle_increment * min_idx;

    float oposite_angle = angle + M_PI;
    // The module of the vector is inverse to the distance to the obstacle
    float complementary_dist = OBSTACLE_DISTANCE - distance_min;

    // Get cartesian (x, y) components from polar (angle, distance)
    vff_vector.repulsive[0] = cos(oposite_angle) * complementary_dist;
    vff_vector.repulsive[1] = sin(oposite_angle) * complementary_dist;
  }

  // Calculate resulting vector adding attractive and repulsive vectors
  vff_vector.result[0] = (vff_vector.repulsive[0] + vff_vector.attractive[0]);
  vff_vector.result[1] = (vff_vector.repulsive[1] + vff_vector.attractive[1]);

  return vff_vector;
}

visualization_msgs::msg::MarkerArray
AvoidanceNode::get_debug_vff(const VFFVectors & vff_vectors)
{
  visualization_msgs::msg::MarkerArray marker_array;
```

br2_vff_avoidance/src/br2_vff_avoidance/AvoidanceNode.cpp

```
  marker_array.markers.push_back(make_marker(vff_vectors.attractive, BLUE));
  marker_array.markers.push_back(make_marker(vff_vectors.repulsive, RED));
  marker_array.markers.push_back(make_marker(vff_vectors.result, GREEN));

  return marker_array;
}

visualization_msgs::msg::Marker
AvoidanceNode::make_marker(const std::vector<float> & vector, VFFColor vff_color)
{
  visualization_msgs::msg::Marker marker;

  marker.header.frame_id = "base_footprint";
  marker.header.stamp = now();
  marker.type = visualization_msgs::msg::Marker::ARROW;
  marker.id = visualization_msgs::msg::Marker::ADD;

  geometry_msgs::msg::Point start;
  start.x = 0.0;
  start.y = 0.0;
  geometry_msgs::msg::Point end;
  start.x = vector[0];
  start.y = vector[1];
  marker.points = {end, start};

  marker.scale.x = 0.05;
  marker.scale.y = 0.1;

  switch (vff_color) {
    case RED:
      marker.id = 0;
      marker.color.r = 1.0;
      break;
    case GREEN:
      marker.id = 1;
      marker.color.g = 1.0;
      break;
    case BLUE:
      marker.id = 2;
      marker.color.b = 1.0;
      break;
  }
  marker.color.a = 1.0;

  return marker;
}

}  // namespace br2_vff_avoidance
```

br2_vff_avoidance/src/avoidance_vff_main.cpp

```
#include <memory>

#include "br2_vff_avoidance/AvoidanceNode.hpp"
#include "rclcpp/rclcpp.hpp"

int main(int argc, char * argv[])
{
  rclcpp::init(argc, argv);

  auto avoidance_node = std::make_shared<br2_vff_avoidance::AvoidanceNode>();
  rclcpp::spin(avoidance_node);

  rclcpp::shutdown();

  return 0;
}
```

br2_vff_avoidance/tests/vff_test.cpp

```
#include <limits>
#include <vector>
#include <memory>

#include "sensor_msgs/msg/laser_scan.hpp"
#include "br2_vff_avoidance/AvoidanceNode.hpp"

#include "gtest/gtest.h"

using namespace std::chrono_literals;

class AvoidanceNodeTest : public br2_vff_avoidance::AvoidanceNode
{
public:
  br2_vff_avoidance::VFFVectors
  get_vff_test(const sensor_msgs::msg::LaserScan & scan)
  {
    return get_vff(scan);
  }

  visualization_msgs::msg::MarkerArray
  get_debug_vff_test(const br2_vff_avoidance::VFFVectors & vff_vectors)
  {
    return get_debug_vff(vff_vectors);
  }
};

sensor_msgs::msg::LaserScan get_scan_test_1(rclcpp::Time ts)
{
  sensor_msgs::msg::LaserScan ret;
  ret.header.stamp = ts;
  ret.angle_min = -M_PI;
  ret.angle_max = M_PI;
  ret.angle_increment = 2.0 * M_PI / 16.0;
  ret.ranges = std::vector<float>(16, std::numeric_limits<float>::infinity());

  return ret;
}

sensor_msgs::msg::LaserScan get_scan_test_2(rclcpp::Time ts)
{
  sensor_msgs::msg::LaserScan ret;
  ret.header.stamp = ts;
  ret.angle_min = -M_PI;
  ret.angle_max = M_PI;
  ret.angle_increment = 2.0 * M_PI / 16.0;
  ret.ranges = std::vector<float>(16, 0.0);

  return ret;
}

sensor_msgs::msg::LaserScan get_scan_test_3(rclcpp::Time ts)
{
  sensor_msgs::msg::LaserScan ret;
  ret.header.stamp = ts;
  ret.angle_min = -M_PI;
  ret.angle_max = M_PI;
  ret.angle_increment = 2.0 * M_PI / 16.0;
  ret.ranges = std::vector<float>(16, 5.0);
  ret.ranges[2] = 0.3;

  return ret;
}

sensor_msgs::msg::LaserScan get_scan_test_4(rclcpp::Time ts)
{
  sensor_msgs::msg::LaserScan ret;
  ret.header.stamp = ts;
  ret.angle_min = -M_PI;
  ret.angle_max = M_PI;
  ret.angle_increment = 2.0 * M_PI / 16.0;
  ret.ranges = std::vector<float>(16, 5.0);
  ret.ranges[6] = 0.3;

  return ret;
}
```

br2_vff_avoidance/tests/vff_test.cpp

```
sensor_msgs::msg::LaserScan get_scan_test_5(rclcpp::Time ts)
{
  sensor_msgs::msg::LaserScan ret;
  ret.header.stamp = ts;
  ret.angle_min = -M_PI;
  ret.angle_max = M_PI;

ret.angle_increment = 2.0 * M_PI / 16.0;
  ret.ranges = std::vector<float>(16, 5.0);
  ret.ranges[10] = 0.3;

  return ret;
}

sensor_msgs::msg::LaserScan get_scan_test_6(rclcpp::Time ts)
{
  sensor_msgs::msg::LaserScan ret;
  ret.header.stamp = ts;
  ret.angle_min = -M_PI;
  ret.angle_max = M_PI;
  ret.angle_increment = 2.0 * M_PI / 16.0;
  ret.ranges = std::vector<float>(16, 0.5);
  ret.ranges[10] = 0.3;

  return ret;
}

sensor_msgs::msg::LaserScan get_scan_test_7(rclcpp::Time ts)
{
  sensor_msgs::msg::LaserScan ret;
  ret.header.stamp = ts;
  ret.angle_min = -M_PI;
  ret.angle_max = M_PI;
  ret.angle_increment = 2.0 * M_PI / 16.0;
  ret.ranges = std::vector<float>(16, 5.0);
  ret.ranges[14] = 0.3;

  return ret;
}

sensor_msgs::msg::LaserScan get_scan_test_8(rclcpp::Time ts)
{
  sensor_msgs::msg::LaserScan ret;
  ret.header.stamp = ts;
  ret.angle_min = -M_PI;
  ret.angle_max = M_PI;
  ret.angle_increment = 2.0 * M_PI / 16.0;
  ret.ranges = std::vector<float>(16, 5.0);
  ret.ranges[8] = 0.01;

  return ret;
}

TEST(vff_tests, get_vff)
{
  auto node_avoidance = AvoidanceNodeTest();

  rclcpp::Time ts = node_avoidance.now();

  auto res1 = node_avoidance.get_vff_test(get_scan_test_1(ts));
  ASSERT_EQ(res1.attractive, std::vector<float>({1.0f, 0.0f}));
  ASSERT_EQ(res1.repulsive, std::vector<float>({0.0f, 0.0f}));
  ASSERT_EQ(res1.result, std::vector<float>({1.0f, 0.0f}));

  auto res2 = node_avoidance.get_vff_test(get_scan_test_2(ts));
  ASSERT_EQ(res2.attractive, std::vector<float>({1.0f, 0.0f}));
  ASSERT_NEAR(res2.repulsive[0], 1.0f, 0.00001f);
  ASSERT_NEAR(res2.repulsive[1], 0.0f, 0.00001f);
  ASSERT_NEAR(res2.result[0], 2.0f, 0.00001f);
  ASSERT_NEAR(res2.result[1], 0.0f, 0.00001f);

  auto res3 = node_avoidance.get_vff_test(get_scan_test_3(ts));
  ASSERT_EQ(res3.attractive, std::vector<float>({1.0f, 0.0f}));
  ASSERT_GT(res3.repulsive[0], 0.0f);
  ASSERT_GT(res3.repulsive[1], 0.0f);
  ASSERT_GT(atan2(res3.repulsive[1], res3.repulsive[0]), 0.1);
  ASSERT_LT(atan2(res3.repulsive[1], res3.repulsive[0]), M_PI_2);
  ASSERT_GT(atan2(res3.result[1], res3.result[0]), 0.1);
  ASSERT_LT(atan2(res3.result[1], res3.result[0]), M_PI_2);
```

br2_vff_avoidance/tests/vff_test.cpp

```
  auto res4 = node_avoidance.get_vff_test(get_scan_test_4(ts));
  ASSERT_EQ(res4.attractive, std::vector<float>({1.0f, 0.0f}));
  ASSERT_LT(res4.repulsive[0], 0.0f);
  ASSERT_GT(res4.repulsive[1], 0.0f);
  ASSERT_GT(atan2(res4.repulsive[1], res4.repulsive[0]), M_PI_2);
  ASSERT_LT(atan2(res4.repulsive[1], res4.repulsive[0]), M_PI);
  ASSERT_GT(atan2(res4.result[1], res4.result[0]), 0.0);
  ASSERT_LT(atan2(res4.result[1], res4.result[0]), M_PI_2);

  auto res5 = node_avoidance.get_vff_test(get_scan_test_5(ts));
  ASSERT_EQ(res5.attractive, std::vector<float>({1.0f, 0.0f}));
  ASSERT_LT(res5.repulsive[0], 0.0f);
  ASSERT_LT(res5.repulsive[1], 0.0f);
  ASSERT_GT(atan2(res5.repulsive[1], res5.repulsive[0]), -M_PI);
  ASSERT_LT(atan2(res5.repulsive[1], res5.repulsive[0]), -M_PI_2);
  ASSERT_LT(atan2(res5.result[1], res5.result[0]), 0.0);
  ASSERT_GT(atan2(res5.result[1], res5.result[0]), -M_PI_2);

  auto res6 = node_avoidance.get_vff_test(get_scan_test_6(ts));
  ASSERT_EQ(res6.attractive, std::vector<float>({1.0f, 0.0f}));
  ASSERT_LT(res6.repulsive[0], 0.0f);
  ASSERT_LT(res6.repulsive[1], 0.0f);
  ASSERT_GT(atan2(res6.repulsive[1], res6.repulsive[0]), -M_PI);
  ASSERT_LT(atan2(res6.repulsive[1], res6.repulsive[0]), -M_PI_2);
  ASSERT_LT(atan2(res6.result[1], res6.result[0]), 0.0);
  ASSERT_GT(atan2(res6.result[1], res6.result[0]), -M_PI_2);

  auto res7 = node_avoidance.get_vff_test(get_scan_test_7(ts));
  ASSERT_EQ(res7.attractive, std::vector<float>({1.0f, 0.0f}));
  ASSERT_GT(res7.repulsive[0], 0.0f);
  ASSERT_LT(res7.repulsive[1], 0.0f);
  ASSERT_LT(atan2(res7.repulsive[1], res7.repulsive[0]), 0.0f);
  ASSERT_GT(atan2(res7.repulsive[1], res7.repulsive[0]), -M_PI_2);
  ASSERT_LT(atan2(res7.result[1], res7.result[0]), 0.0);
  ASSERT_GT(atan2(res7.result[1], res7.result[0]), -M_PI_2);

  auto res8 = node_avoidance.get_vff_test(get_scan_test_8(ts));
  ASSERT_EQ(res8.attractive, std::vector<float>({1.0f, 0.0f}));
  ASSERT_NEAR(res8.repulsive[0], -1.0f, 0.1f);
  ASSERT_NEAR(res8.repulsive[1], 0.0f, 0.0001f);
  ASSERT_NEAR(res8.result[0], 0.0f, 0.01f);
  ASSERT_NEAR(res8.result[1], 0.0f, 0.01f);
}

TEST(vff_tests, ouput_vels)
{
  auto node_avoidance = std::make_shared<AvoidanceNodeTest>();

  // Create a testing node with a scan publisher and a speed subscriber
  auto test_node = rclcpp::Node::make_shared("test_node");
  auto scan_pub = test_node->create_publisher<sensor_msgs::msg::LaserScan>(
    "input_scan", 100);

  geometry_msgs::msg::Twist last_vel;
  auto vel_sub = test_node->create_subscription<geometry_msgs::msg::Twist>(
    "output_vel", 1, [&last_vel](geometry_msgs::msg::Twist::SharedPtr msg) {
      last_vel = *msg;
    });

  ASSERT_EQ(vel_sub->get_publisher_count(), 1);
  ASSERT_EQ(scan_pub->get_subscription_count(), 1);

  rclcpp::Rate rate(30);
  rclcpp::executors::SingleThreadedExecutor executor;
  executor.add_node(node_avoidance);
  executor.add_node(test_node);

  // Test for scan test #1
  auto start = node_avoidance->now();
  while (rclcpp::ok() && (node_avoidance->now() - start) < 1s) {
    scan_pub->publish(get_scan_test_1(node_avoidance->now()));
    executor.spin_some();
    rate.sleep();
  }
  ASSERT_NEAR(last_vel.linear.x, 0.3f, 0.0001f);
  ASSERT_NEAR(last_vel.angular.z, 0.0f, 0.0001f);
```

br2_vff_avoidance/tests/vff_test.cpp

```
// Test for scan test #2
start = node_avoidance->now();
while (rclcpp::ok() && (node_avoidance->now() - start) < 1s) {
  scan_pub->publish(get_scan_test_2(node_avoidance->now()));
  executor.spin_some();
  rate.sleep();
}
ASSERT_NEAR(last_vel.linear.x, 0.3f, 0.0001f);
ASSERT_NEAR(last_vel.angular.z, 0.0f, 0.0001f);

// Test for scan test #3
start = node_avoidance->now();
while (rclcpp::ok() && (node_avoidance->now() - start) < 1s) {
  scan_pub->publish(get_scan_test_3(node_avoidance->now()));
  executor.spin_some();
  rate.sleep();
}
ASSERT_LT(last_vel.linear.x, 0.3f);
ASSERT_GT(last_vel.linear.x, 0.0f);
ASSERT_GT(last_vel.angular.z, 0.0f);
ASSERT_LT(last_vel.angular.z, M_PI_2);

// Test for scan test #4
start = node_avoidance->now();
while (rclcpp::ok() && (node_avoidance->now() - start) < 1s) {
  scan_pub->publish(get_scan_test_4(node_avoidance->now()));
  executor.spin_some();
  rate.sleep();
}
ASSERT_LT(last_vel.linear.x, 0.3f);
ASSERT_GT(last_vel.linear.x, 0.0f);
ASSERT_GT(last_vel.angular.z, 0.0f);
ASSERT_LT(last_vel.angular.z, M_PI_2);

// Test for scan test #5
start = node_avoidance->now();
while (rclcpp::ok() && (node_avoidance->now() - start) < 1s) {
  scan_pub->publish(get_scan_test_5(node_avoidance->now()));
  executor.spin_some();
  rate.sleep();
}
ASSERT_LT(last_vel.linear.x, 0.3f);
ASSERT_GT(last_vel.linear.x, 0.0f);
ASSERT_LT(last_vel.angular.z, 0.0f);
ASSERT_GT(last_vel.angular.z, -M_PI_2);

// Test for scan test #6
start = node_avoidance->now();
while (rclcpp::ok() && (node_avoidance->now() - start) < 1s) {
  scan_pub->publish(get_scan_test_6(node_avoidance->now()));
  executor.spin_some();
  rate.sleep();
}
ASSERT_LT(last_vel.linear.x, 0.3f);
ASSERT_GT(last_vel.linear.x, 0.0f);
ASSERT_LT(last_vel.angular.z, 0.0f);
ASSERT_GT(last_vel.angular.z, -M_PI_2);

// Test for scan test #7
start = node_avoidance->now();
while (rclcpp::ok() && (node_avoidance->now() - start) < 1s) {
  scan_pub->publish(get_scan_test_7(node_avoidance->now()));
  executor.spin_some();
  rate.sleep();
}
ASSERT_LT(last_vel.linear.x, 0.3f);
ASSERT_GT(last_vel.linear.x, 0.0f);
ASSERT_LT(last_vel.angular.z, 0.0f);
ASSERT_GT(last_vel.angular.z, -M_PI_2);
```

br2_vff_avoidance/tests/vff_test.cpp

```
// Test for scan test #8
  start = node_avoidance->now();
  while (rclcpp::ok() && (node_avoidance->now() - start) < 2s) {
    scan_pub->publish(get_scan_test_8(node_avoidance->now()));
    executor.spin_some();
    rate.sleep();
  }
  ASSERT_NEAR(last_vel.linear.x, 0.0f, 0.1f);
  ASSERT_LT(last_vel.angular.z, 0.0f);
  ASSERT_GT(last_vel.angular.z, -M_PI_2);

  // Test for stooping when scan is too old
  last_vel = geometry_msgs::msg::Twist();
  while (rclcpp::ok() && (node_avoidance->now() - start) < 3s) {
    scan_pub->publish(get_scan_test_6(start));
    executor.spin_some();
    rate.sleep();
  }
  ASSERT_NEAR(last_vel.linear.x, 0.0f, 0.01f);
  ASSERT_NEAR(last_vel.angular.z, 0.0f, 0.01f);
}

int main(int argc, char ** argv)
{
  rclcpp::init(argc, argv);

  testing::InitGoogleTest(&argc, argv);
  return RUN_ALL_TESTS();
}
```

br2_vff_avoidance/tests/CMakeLists.txt

```
ament_add_gtest(vff_test vff_test.cpp)
ament_target_dependencies(vff_test ${dependencies})
target_link_libraries(vff_test ${PROJECT_NAME})
```

A.6 PACKAGE BR2_TRACKING_MSGS

Package br2_tracking_msgs

```
br2_tracking_msgs
├── CMakeLists.txt
├── msg
│   └── PanTiltCommand.msg
```

br2_tracking_msgs/CMakeLists.txt

```
project(br2_tracking_msgs)

cmake_minimum_required(VERSION 3.5)

find_package(ament_cmake REQUIRED)
find_package(builtin_interfaces REQUIRED)
find_package(rosidl_default_generators REQUIRED)

rosidl_generate_interfaces(${PROJECT_NAME}
  "msg/PanTiltCommand.msg"
  DEPENDENCIES builtin_interfaces
)

ament_export_dependencies(rosidl_default_runtime)
ament_package()
```

br2_tracking_msgs/package.xml

```
<?xml version="1.0"?>
<?xml-model href="http://download.ros.org/schema/package_format3.xsd"
  schematypens="http://www.w3.org/2001/XMLSchema"?>
<package format="3">
  <name>br2_tracking_msgs</name>
  <version>0.0.0</version>

  <description>Messages for br2_tracking</description>

  <maintainer email="fmrico@gmail.com">Francisco Martín</maintainer>

  <license>Apache 2.0</license>

  <buildtool_depend>ament_cmake</buildtool_depend>

  <depend>rclcpp</depend>
  <depend>builtin_interfaces</depend>
  <depend>rosidl_default_generators</depend>

  <member_of_group>rosidl_interface_packages</member_of_group>

  <export>
    <build_type>ament_cmake</build_type>
  </export>
</package>
```

A.7 PACKAGE BR2_TRACKING

Package br2_tracking

```
br2_tracking
├── CMakeLists.txt
├── config
│   └── detector.yaml
├── include
│   └── br2_tracking
│       ├── HeadController.hpp
│       ├── ObjectDetector.hpp
│       └── PIDController.hpp
├── launch
│   └── tracking.launch.py
├── package.xml
├── src
│   ├── br2_tracking
│   │   ├── HeadController.cpp
│   │   ├── ObjectDetector.cpp
│   │   └── PIDController.cpp
│   ├── object_detector_main.cpp
│   └── object_tracker_main.cpp
└── tests
    ├── CMakeLists.txt
```

br2_tracking/CMakeLists.txt

```
cmake_minimum_required(VERSION 3.5)
project(br2_tracking)

set(CMAKE_CXX_STANDARD 17)
set(CMAKE_BUILD_TYPE Debug)

find_package(ament_cmake REQUIRED)
find_package(rclcpp REQUIRED)
find_package(rclcpp_lifecycle REQUIRED)
find_package(br2_tracking_msgs REQUIRED)
find_package(sensor_msgs REQUIRED)
find_package(geometry_msgs REQUIRED)
find_package(vision_msgs REQUIRED)
find_package(control_msgs REQUIRED)
find_package(image_transport REQUIRED)
find_package(cv_bridge REQUIRED)

find_package(OpenCV REQUIRED)

set(dependencies
  rclcpp
  rclcpp_lifecycle
  br2_tracking_msgs
  sensor_msgs
  geometry_msgs
  vision_msgs
  control_msgs
  image_transport
  cv_bridge
  OpenCV
)

include_directories(include)

add_library(${PROJECT_NAME} SHARED
  src/br2_tracking/ObjectDetector.cpp
  src/br2_tracking/HeadController.cpp
  src/br2_tracking/PIDController.cpp
)
ament_target_dependencies(${PROJECT_NAME} ${dependencies})

add_executable(object_detector src/object_detector_main.cpp)
ament_target_dependencies(object_detector ${dependencies})
target_link_libraries(object_detector ${PROJECT_NAME})

add_executable(object_tracker src/object_tracker_main.cpp)
ament_target_dependencies(object_tracker ${dependencies})
target_link_libraries(object_tracker ${PROJECT_NAME})


install(TARGETS
  ${PROJECT_NAME}
  object_detector
  object_tracker
  ARCHIVE DESTINATION lib
  LIBRARY DESTINATION lib
  RUNTIME DESTINATION lib/${PROJECT_NAME}
)

install(
  DIRECTORY include
  DESTINATION include
)

install(DIRECTORY launch config DESTINATION share/${PROJECT_NAME})

if(BUILD_TESTING)
  find_package(ament_lint_auto REQUIRED)
  ament_lint_auto_find_test_dependencies()

  set(ament_cmake_cpplint_FOUND TRUE)
  ament_lint_auto_find_test_dependencies()

  find_package(ament_cmake_gtest REQUIRED)
  add_subdirectory(tests)
endif()
```

br2_tracking/CMakeLists.txt

```
ament_export_include_directories(include)
ament_export_libraries(${PROJECT_NAME})
ament_export_dependencies(${dependencies})
ament_package()
```

br2_tracking/launch/tracking.launch.py

```
import os

from ament_index_python.packages import get_package_share_directory

from launch import LaunchDescription
from launch_ros.actions import Node


def generate_launch_description():

    params_file = os.path.join(
        get_package_share_directory('br2_tracking'),
        'config',
        'detector.yaml'
        )

    object_tracker_cmd = Node(
        package='br2_tracking',
        executable='object_tracker',
        parameters=[{
          'use_sim_time': True
        }, params_file],
        remappings=[
          ('input_image', '/head_front_camera/rgb/image_raw'),
          ('joint_state', '/head_controller/state'),
          ('joint_command', '/head_controller/joint_trajectory')
        ],
        output='screen'
    )

    ld = LaunchDescription()

    # Add any actions
    ld.add_action(object_tracker_cmd)

    return ld
```

br2_tracking/package.xml

```
<?xml version="1.0"?>
<?xml-model href="http://download.ros.org/schema/package_format3.xsd"
  schematypens="http://www.w3.org/2001/XMLSchema"?>
<package format="3">
  <name>br2_tracking</name>
  <version>0.0.0</version>
  <description>Tracking package</description>
  <maintainer email="fmrico@gmail.com">Francisco Martín</maintainer>
  <license>Apache 2.0</license>

  <buildtool_depend>ament_cmake</buildtool_depend>

  <depend>rclcpp</depend>
  <depend>rclcpp_lifecycle</depend>
  <depend>geometry_msgs</depend>
  <depend>br2_tracking_msgs</depend>
  <depend>sensor_msgs</depend>
  <depend>vision_msgs</depend>
  <depend>control_msgs</depend>
  <depend>image_transport</depend>
  <depend>cv_bridge</depend>

  <test_depend>ament_lint_auto</test_depend>
  <test_depend>ament_lint_common</test_depend>
  <test_depend>ament_cmake_gtest</test_depend>

  <export>
    <build_type>ament_cmake</build_type>
  </export>
</package>
```

br2_tracking/include/br2_tracking/PIDController.hpp

```
#ifndef BR2_TRACKING__PIDCONTROLLER_HPP_
#define BR2_TRACKING__PIDCONTROLLER_HPP_

#include <cmath>

namespace br2_tracking
{

class PIDController
{
public:
  PIDController(double min_ref, double max_ref, double min_output, double max_output);

  void set_pid(double n_KP, double n_KI, double n_KD);
  double get_output(double new_reference);

private:
  double KP_, KI_, KD_;

  double min_ref_, max_ref_;
  double min_output_, max_output_;
  double prev_error_, int_error_;
};

}  // namespace br2_tracking

#endif  // BR2_TRACKING__PIDCONTROLLER_HPP_
```

br2_tracking/include/br2_tracking/ObjectDetector.hpp

```
#ifndef BR2_TRACKING__OBJECTDETECTOR_HPP_
#define BR2_TRACKING__OBJECTDETECTOR_HPP_

#include <memory>
#include <vector>

#include "vision_msgs/msg/detection2_d.hpp"

#include "image_transport/image_transport.hpp"
#include "rclcpp/rclcpp.hpp"

namespace br2_tracking
{

class ObjectDetector : public rclcpp::Node
{
public:
  ObjectDetector();

  void image_callback(const sensor_msgs::msg::Image::ConstSharedPtr & msg);

private:
  image_transport::Subscriber image_sub_;
  rclcpp::Publisher<vision_msgs::msg::Detection2D>::SharedPtr detection_pub_;

  // HSV ranges for detection [h - H] [s - S] [v - V]
  std::vector<double> hsv_filter_ranges_ {0, 180, 0, 255, 0, 255};
  bool debug_ {true};
};

}  // namespace br2_tracking

#endif  // BR2_TRACKING__OBJECTDETECTOR_HPP_
```

br2_tracking/include/br2_tracking/HeadController.hpp

```
#ifndef BR2_TRACKING__HEADCONTROLLER_HPP_
#define BR2_TRACKING__HEADCONTROLLER_HPP_

#include <memory>

#include "br2_tracking_msgs/msg/pan_tilt_command.hpp"
#include "control_msgs/msg/joint_trajectory_controller_state.hpp"
#include "trajectory_msgs/msg/joint_trajectory.hpp"

#include "br2_tracking/PIDController.hpp"

#include "image_transport/image_transport.hpp"

#include "rclcpp_lifecycle/lifecycle_node.hpp"
#include "rclcpp/rclcpp.hpp"

namespace br2_tracking
{

using CallbackReturn =
  rclcpp_lifecycle::node_interfaces::LifecycleNodeInterface::CallbackReturn;

class HeadController : public rclcpp_lifecycle::LifecycleNode
{
public:
  HeadController();

  CallbackReturn on_configure(const rclcpp_lifecycle::State & previous_state);
  CallbackReturn on_activate(const rclcpp_lifecycle::State & previous_state);
  CallbackReturn on_deactivate(const rclcpp_lifecycle::State & previous_state);

  void control_sycle();

  void joint_state_callback(
    control_msgs::msg::JointTrajectoryControllerState::UniquePtr msg);
  void command_callback(br2_tracking_msgs::msg::PanTiltCommand::UniquePtr msg);

private:
  rclcpp::Subscription<br2_tracking_msgs::msg::PanTiltCommand>::SharedPtr command_sub_;
  rclcpp::Subscription<control_msgs::msg::JointTrajectoryControllerState>::SharedPtr
    joint_sub_;
  rclcpp_lifecycle::LifecyclePublisher<trajectory_msgs::msg::JointTrajectory>::SharedPtr
    joint_pub_;
  rclcpp::TimerBase::SharedPtr timer_;

  control_msgs::msg::JointTrajectoryControllerState::UniquePtr last_state_;
  br2_tracking_msgs::msg::PanTiltCommand::UniquePtr last_command_;
  rclcpp::Time last_command_ts_;

  PIDController pan_pid_, tilt_pid_;
};

}  // namespace br2_tracking

#endif  // BR2_TRACKING__HEADCONTROLLER_HPP_
```

br2_tracking/config/detector.yaml

```
/object_detector:
  ros__parameters:
    debug: true
    hsv_ranges:
      - 15.0
      - 20.0
      - 50.0
      - 200.0
      - 20.0
      - 200.0
```

br2_tracking/src/br2_tracking/PIDController.cpp

```
#include <algorithm>

#include "br2_tracking/PIDController.hpp"

namespace br2_tracking
{

PIDController::PIDController(
  double min_ref, double max_ref, double min_output, double max_output)
{
  min_ref_ = min_ref;
  max_ref_ = max_ref;
  min_output_ = min_output;
  max_output_ = max_output;
  prev_error_ = int_error_ = 0.0;

  KP_ = 0.41;
  KI_ = 0.06;
  KD_ = 0.53;
}

void
PIDController::set_pid(double n_KP, double n_KI, double n_KD)
{
  KP_ = n_KP;
  KI_ = n_KI;
  KD_ = n_KD;
}

double
PIDController::get_output(double new_reference)
{
  double ref = new_reference;
  double output = 0.0;

  // Proportional Error
  double direction = 0.0;
  if (ref != 0.0) {
    direction = ref / fabs(ref);
  }

  if (fabs(ref) < min_ref_) {
    output = 0.0;
  } else if (fabs(ref) > max_ref_) {
    output = direction * max_output_;
  } else {
    output = direction * min_output_ + ref * (max_output_ - min_output_);
  }

  // Integral Error
  int_error_ = (int_error_ + output) * 2.0 / 3.0;

  // Derivative Error
  double deriv_error = output - prev_error_;
  prev_error_ = output;

  output = KP_ * output + KI_ * int_error_ + KD_ * deriv_error;

  return std::clamp(output, -max_output_, max_output_);
}

}  // namespace br2_tracking
```

br2_tracking/src/br2_tracking/HeadController.cpp

```
#include <algorithm>
#include <utility>

#include "br2_tracking/HeadController.hpp"
#include "br2_tracking/PIDController.hpp"

#include "br2_tracking_msgs/msg/pan_tilt_command.hpp"
#include "control_msgs/msg/joint_trajectory_controller_state.hpp"
#include "trajectory_msgs/msg/joint_trajectory.hpp"

#include "rclcpp_lifecycle/lifecycle_node.hpp"
#include "rclcpp/rclcpp.hpp"

namespace br2_tracking
{

using std::placeholders::_1;
using namespace std::chrono_literals;
using CallbackReturn = rclcpp_lifecycle::node_interfaces::LifecycleNodeInterface::
  CallbackReturn;

HeadController::HeadController()
: LifecycleNode("head_tracker"),
  pan_pid_(0.0, 1.0, 0.0, 0.3),
  tilt_pid_(0.0, 1.0, 0.0, 0.3)
{
  command_sub_ = create_subscription<br2_tracking_msgs::msg::PanTiltCommand>(
    "command", 100,
    std::bind(&HeadController::command_callback, this, _1));
  joint_sub_ = create_subscription<control_msgs::msg::JointTrajectoryControllerState>(
    "joint_state", rclcpp::SensorDataQoS(),
    std::bind(&HeadController::joint_state_callback, this, _1));
  joint_pub_ = create_publisher<trajectory_msgs::msg::JointTrajectory>("joint_command",
   10);
}

CallbackReturn
HeadController::on_configure(const rclcpp_lifecycle::State & previous_state)
{
  RCLCPP_INFO(get_logger(), "HeadController configured");

  pan_pid_.set_pid(0.4, 0.05, 0.55);
  tilt_pid_.set_pid(0.4, 0.05, 0.55);

  return CallbackReturn::SUCCESS;
}

CallbackReturn
HeadController::on_activate(const rclcpp_lifecycle::State & previous_state)
{
  RCLCPP_INFO(get_logger(), "HeadController activated");

  joint_pub_->on_activate();
  timer_ = create_wall_timer(100ms, std::bind(&HeadController::control_sycle, this));

  return CallbackReturn::SUCCESS;
}

CallbackReturn
HeadController::on_deactivate(const rclcpp_lifecycle::State & previous_state)
{
  RCLCPP_INFO(get_logger(), "HeadController deactivated");

  trajectory_msgs::msg::JointTrajectory command_msg;
  command_msg.header.stamp = now();
  command_msg.joint_names = last_state_->joint_names;
  command_msg.points.resize(1);
  command_msg.points[0].positions.resize(2);
  command_msg.points[0].velocities.resize(2);
  command_msg.points[0].accelerations.resize(2);
  command_msg.points[0].positions[0] = 0.0;
  command_msg.points[0].positions[1] = 0.0;
  command_msg.points[0].velocities[0] = 0.1;
  command_msg.points[0].velocities[1] = 0.1;
  command_msg.points[0].accelerations[0] = 0.1;
  command_msg.points[0].accelerations[1] = 0.1;
  command_msg.points[0].time_from_start = rclcpp::Duration(1s);
```

br2_tracking/src/br2_tracking/HeadController.cpp

```
 joint_pub_->publish(command_msg);

  joint_pub_->on_deactivate();
  timer_ = nullptr;

  return CallbackReturn::SUCCESS;
}

void
HeadController::joint_state_callback(
  control_msgs::msg::JointTrajectoryControllerState::UniquePtr msg)
{
  last_state_ = std::move(msg);
}

void
HeadController::command_callback(br2_tracking_msgs::msg::PanTiltCommand::UniquePtr msg)
{
  last_command_ = std::move(msg);
  last_command_ts_ = now();
}

void
HeadController::control_sycle()
{
  if (last_state_ == nullptr) {return;}

  trajectory_msgs::msg::JointTrajectory command_msg;
  command_msg.header.stamp = now();
  command_msg.joint_names = last_state_->joint_names;
  command_msg.points.resize(1);
  command_msg.points[0].positions.resize(2);
  command_msg.points[0].velocities.resize(2);
  command_msg.points[0].accelerations.resize(2);
  command_msg.points[0].time_from_start = rclcpp::Duration(200ms);

  if (last_command_ == nullptr || (now() - last_command_ts_) > 100ms) {
    command_msg.points[0].positions[0] = 0.0;
    command_msg.points[0].positions[1] = 0.0;
    command_msg.points[0].velocities[0] = 0.1;
    command_msg.points[0].velocities[1] = 0.1;
    command_msg.points[0].accelerations[0] = 0.1;
    command_msg.points[0].accelerations[1] = 0.1;
    command_msg.points[0].time_from_start = rclcpp::Duration(1s);
  } else {
    double control_pan = pan_pid_.get_output(last_command_->pan);
    double control_tilt = tilt_pid_.get_output(last_command_->tilt);

    command_msg.points[0].positions[0] = last_state_->actual.positions[0] - control_pan;
    command_msg.points[0].positions[1] = last_state_->actual.positions[1] - control_tilt;

    command_msg.points[0].velocities[0] = 0.5;
    command_msg.points[0].velocities[1] = 0.5;
    command_msg.points[0].accelerations[0] = 0.5;
    command_msg.points[0].accelerations[1] = 0.5;
  }

  joint_pub_->publish(command_msg);
}

}  // namespace br2_tracking
```

br2_tracking/src/br2_tracking/ObjectDetector.cpp

```
#include <vector>

#include "opencv2/opencv.hpp"
#include "cv_bridge/cv_bridge.h"

#include "br2_tracking/ObjectDetector.hpp"
#include "geometry_msgs/msg/pose2_d.hpp"

#include "image_transport/image_transport.hpp"
#include "rclcpp/rclcpp.hpp"

namespace br2_tracking
{
using std::placeholders::_1;

ObjectDetector::ObjectDetector()
: Node("object_detector")
{
  image_sub_ = image_transport::create_subscription(
    this, "input_image", std::bind(&ObjectDetector::image_callback, this, _1),
    "raw", rclcpp::SensorDataQoS().get_rmw_qos_profile());

  detection_pub_ = create_publisher<vision_msgs::msg::Detection2D>("detection", 100);

  declare_parameter("hsv_ranges", hsv_filter_ranges_);
  declare_parameter("debug", debug_);
  get_parameter("hsv_ranges", hsv_filter_ranges_);
  get_parameter("debug", debug_);
}

void
ObjectDetector::image_callback(const sensor_msgs::msg::Image::ConstSharedPtr & msg)
{
  if (detection_pub_->get_subscription_count() == 0) {return;}

  const float & h = hsv_filter_ranges_[0];
  const float & H = hsv_filter_ranges_[1];
  const float & s = hsv_filter_ranges_[2];
  const float & S = hsv_filter_ranges_[3];
  const float & v = hsv_filter_ranges_[4];
  const float & V = hsv_filter_ranges_[5];

  cv_bridge::CvImagePtr cv_ptr;
  try {
    cv_ptr = cv_bridge::toCvCopy(msg, sensor_msgs::image_encodings::BGR8);
  } catch (cv_bridge::Exception & e) {
    RCLCPP_ERROR(get_logger(), "cv_bridge exception: %s", e.what());
    return;
  }

  cv::Mat img_hsv;
  cv::cvtColor(cv_ptr->image, img_hsv, cv::COLOR_BGR2HSV);

  cv::Mat1b filtered;
  cv::inRange(img_hsv, cv::Scalar(h, s, v), cv::Scalar(H, S, V), filtered);

  auto moment = cv::moments(filtered, true);
  cv::Rect bbx = cv::boundingRect(filtered);

  auto m = cv::moments(filtered, true);
  if (m.m00 < 0.000001) {return;}
  int cx = m.m10 / m.m00;
  int cy = m.m01 / m.m00;

  vision_msgs::msg::Detection2D detection_msg;
  detection_msg.header = msg->header;
  detection_msg.bbox.size_x = bbx.width;
  detection_msg.bbox.size_y = bbx.height;
  detection_msg.bbox.center.x = cx;
  detection_msg.bbox.center.y = cy;
  detection_msg.source_img = *cv_ptr->toImageMsg();
  detection_pub_->publish(detection_msg);

  if (debug_) {
    cv::rectangle(cv_ptr->image, bbx, cv::Scalar(0, 0, 255), 3);
    cv::circle(cv_ptr->image, cv::Point(cx, cy), 3, cv::Scalar(255, 0, 0), 3);
    cv::imshow("cv_ptr->image", cv_ptr->image);
```

br2_tracking/src/br2_tracking/ObjectDetector.cpp

```
    cv::waitKey(1);
  }
}

}  // namespace br2_tracking
```

br2_tracking/src/object_tracker_main.cpp

```
#include <memory>

#include "br2_tracking/ObjectDetector.hpp"
#include "br2_tracking/HeadController.hpp"

#include "br2_tracking_msgs/msg/pan_tilt_command.hpp"

#include "lifecycle_msgs/msg/transition.hpp"
#include "rclcpp/rclcpp.hpp"

int main(int argc, char * argv[])
{
  rclcpp::init(argc, argv);

  auto node_detector = std::make_shared<br2_tracking::ObjectDetector>();
  auto node_head_controller = std::make_shared<br2_tracking::HeadController>();
  auto node_tracker = rclcpp::Node::make_shared("tracker");

  auto command_pub =
  node_tracker->create_publisher<br2_tracking_msgs::msg::PanTiltCommand>("/command", 100);
  auto detection_sub = node_tracker->create_subscription<vision_msgs::msg::Detection2D>(
    "/detection", rclcpp::SensorDataQoS(),
    [command_pub](vision_msgs::msg::Detection2D::SharedPtr msg) {
      br2_tracking_msgs::msg::PanTiltCommand command;
      command.pan = (msg->bbox.center.x / msg->source_img.width) * 2.0 - 1.0;
      command.tilt = (msg->bbox.center.y / msg->source_img.height) * 2.0 - 1.0;
      command_pub->publish(command);
    });

  rclcpp::executors::SingleThreadedExecutor executor;
  executor.add_node(node_detector);
  executor.add_node(node_head_controller->get_node_base_interface());
  executor.add_node(node_tracker);

  node_head_controller->trigger_transition(
    lifecycle_msgs::msg::Transition::TRANSITION_CONFIGURE);

  executor.spin();

  rclcpp::shutdown();
  return 0;
}
```

br2_tracking/src/object_detector_main.cpp

```
#include <memory>

#include "br2_tracking/ObjectDetector.hpp"
#include "rclcpp/rclcpp.hpp"

int main(int argc, char * argv[])
{
  rclcpp::init(argc, argv);

  auto node_detector = std::make_shared<br2_tracking::ObjectDetector>();

  rclcpp::spin(node_detector);

  rclcpp::shutdown();
  return 0;
}
```

br2_tracking/tests/CMakeLists.txt

```
ament_add_gtest(pid_test pid_test.cpp)
ament_target_dependencies(pid_test ${dependencies})
target_link_libraries(pid_test ${PROJECT_NAME})
```

br2_tracking/tests/pid_test.cpp

```
#include <random>

#include "br2_tracking/PIDController.hpp"

#include "gtest/gtest.h"

TEST(pid_tests, pid_test_1)
{
  br2_tracking::PIDController pid(0.0, 1.0, 0.0, 1.0);

  ASSERT_NEAR(pid.get_output(0.0), 0.0, 0.05);
  ASSERT_LT(pid.get_output(0.1), 0.099);
  ASSERT_GT(pid.get_output(0.1), -0.4);
  ASSERT_LT(pid.get_output(0.1), 0.3);
}

TEST(pid_tests, pid_test_2)
{
  br2_tracking::PIDController pid(0.0, 1.0, 0.0, 1.0);
  pid.set_pid(1.0, 0.0, 0.0);

  std::random_device rd;
  std::mt19937 gen(rd());
  std::uniform_real_distribution<> dis(-5.0, 5.0);

  for (int n = 0; n < 100000; n++) {
    double random_number = dis(gen);
    double output = pid.get_output(random_number);

    ASSERT_LE(output, 1.0);
    ASSERT_GE(output, -1.0);

    if (output < -2.0) {
      ASSERT_NEAR(output, -1.0, 0.01);
    }
    if (output > 2.0) {
      ASSERT_NEAR(output, 1.0, 0.01);
    }
    if (output > 0.0) {
      ASSERT_GT(output, 0.0);
    }
    if (output < 0.0) {
      ASSERT_LT(output, 0.0);
    }
  }
}

int main(int argc, char ** argv)
{
  testing::InitGoogleTest(&argc, argv);
  return RUN_ALL_TESTS();
}
```

A.8 PACKAGE BR2_BT_BUMPGO

Package br2_bt_bumpgo

```
br2_bt_bumpgo
├── behavior_tree_xml
│   └── bumpgo.xml
├── cmake
│   └── FindZMQ.cmake
├── CMakeLists.txt
├── include
│   └── br2_bt_bumpgo
│       ├── Back.hpp
│       ├── Forward.hpp
│       ├── IsObstacle.hpp
│       └── Turn.hpp
├── package.xml
├── src
│   ├── br2_bt_bumpgo
│   │   ├── Back.cpp
│   │   ├── Forward.cpp
│   │   ├── IsObstacle.cpp
│   │   └── Turn.cpp
│   └── bt_bumpgo_main.cpp
└── tests
    ├── bt_action_test.cpp
    ├── bt_forward_main.cpp
```

br2_bt_bumpgo/behavior_tree_xml/bumpgo.xml

```
<?xml version="1.0"?>
<root main_tree_to_execute="BehaviorTree">
    <!-- ////////// -->
    <BehaviorTree ID="BehaviorTree">
        <ReactiveSequence>
            <Fallback>
                <Inverter>
                    <Condition ID="IsObstacle" distance="1.0"/>
                </Inverter>
                <Sequence>
                    <Action ID="Back"/>
                    <Action ID="Turn"/>
                </Sequence>
            </Fallback>
            <Action ID="Forward"/>
        </ReactiveSequence>
    </BehaviorTree>
    <!-- ////////// -->
    <TreeNodesModel>
        <Action ID="Back"/>
        <Action ID="Forward"/>
        <Condition ID="IsObstacle">
            <input_port default="1.0" name="distance">
              Distance to consider obstacle
            </input_port>
        </Condition>
        <Action ID="Turn"/>
    </TreeNodesModel>
    <!-- ////////// -->
</root>
```

br2_bt_bumpgo/CMakeLists.txt

```
cmake_minimum_required(VERSION 3.5)
project(br2_bt_bumpgo)

set(CMAKE_CONFIG_PATH ${CMAKE_MODULE_PATH}  "${CMAKE_CURRENT_LIST_DIR}/cmake")
list(APPEND CMAKE_MODULE_PATH "${CMAKE_CONFIG_PATH}")

find_package(ament_cmake REQUIRED)
find_package(rclcpp REQUIRED)
find_package(behaviortree_cpp_v3 REQUIRED)
find_package(sensor_msgs REQUIRED)
find_package(geometry_msgs REQUIRED)
find_package(ament_index_cpp REQUIRED)

find_package(ZMQ)
if(ZMQ_FOUND)
    message(STATUS "ZeroMQ found.")
    add_definitions(-DZMQ_FOUND)
else()
  message(WARNING "ZeroMQ NOT found. Not including PublisherZMQ.")
endif()

set(CMAKE_CXX_STANDARD 17)

set(dependencies
    rclcpp
    behaviortree_cpp_v3
    sensor_msgs
    geometry_msgs
    ament_index_cpp
)

include_directories(include ${ZMQ_INCLUDE_DIRS})

add_library(br2_forward_bt_node SHARED src/br2_bt_bumpgo/Forward.cpp)
add_library(br2_back_bt_node SHARED src/br2_bt_bumpgo/Back.cpp)
add_library(br2_turn_bt_node SHARED src/br2_bt_bumpgo/Turn.cpp)
add_library(br2_is_obstacle_bt_node SHARED src/br2_bt_bumpgo/IsObstacle.cpp)

list(APPEND plugin_libs
  br2_forward_bt_node
  br2_back_bt_node
  br2_turn_bt_node
  br2_is_obstacle_bt_node
)

foreach(bt_plugin ${plugin_libs})
  ament_target_dependencies(${bt_plugin} ${dependencies})
  target_compile_definitions(${bt_plugin} PRIVATE BT_PLUGIN_EXPORT)
endforeach()

add_executable(bt_bumpgo src/bt_bumpgo_main.cpp)
ament_target_dependencies(bt_bumpgo ${dependencies})
target_link_libraries(bt_bumpgo ${ZMQ_LIBRARIES})

install(TARGETS
  ${plugin_libs}
  bt_bumpgo
  ARCHIVE DESTINATION lib
  LIBRARY DESTINATION lib
  RUNTIME DESTINATION lib/${PROJECT_NAME}
)

install(DIRECTORY include/
  DESTINATION include/
)

install(DIRECTORY behavior_tree_xml
  DESTINATION share/${PROJECT_NAME}
)


if(BUILD_TESTING)
  find_package(ament_lint_auto REQUIRED)
  ament_lint_auto_find_test_dependencies()

  set(ament_cmake_cpplint_FOUND TRUE)
  ament_lint_auto_find_test_dependencies()
```

br2_bt_bumpgo/CMakeLists.txt

```
  find_package(ament_cmake_gtest REQUIRED)

  add_subdirectory(tests)
endif()

ament_export_include_directories(include)
ament_export_dependencies(${dependencies})

ament_package()
```

br2_bt_bumpgo/package.xml

```
<?xml version="1.0"?>
<?xml-model href="http://download.ros.org/schema/package_format3.xsd"
  schematypens="http://www.w3.org/2001/XMLSchema"?>
<package format="3">
  <name>br2_bt_bumpgo</name>
  <version>0.0.0</version>
  <description>BumpGo with Bhehavior Trees package</description>
  <maintainer email="fmrico@gmail.com">fmrico</maintainer>
  <license>Apache 2.0</license>

  <buildtool_depend>ament_cmake</buildtool_depend>

  <depend>rclcpp</depend>
  <depend>behaviortree_cpp_v3</depend>
  <depend>sensor_msgs</depend>
  <depend>geometry_msgs</depend>
  <depend>libzmq3-dev</depend>
  <depend>ament_index_cpp</depend>

  <test_depend>ament_lint_auto</test_depend>
  <test_depend>ament_lint_common</test_depend>
  <test_depend>ament_cmake_gtest</test_depend>

  <export>
    <build_type>ament_cmake</build_type>
  </export>
</package>
```

br2_bt_bumpgo/include/br2_bt_bumpgo/Turn.hpp

```
#ifndef BR2_BT_BUMPGO__TURN_HPP_
#define BR2_BT_BUMPGO__TURN_HPP_

#include <string>

#include "behaviortree_cpp_v3/behavior_tree.h"
#include "behaviortree_cpp_v3/bt_factory.h"

#include "geometry_msgs/msg/twist.hpp"
#include "rclcpp/rclcpp.hpp"

namespace br2_bt_bumpgo
{

class Turn : public BT::ActionNodeBase
{
public:
  explicit Turn(
    const std::string & xml_tag_name,
    const BT::NodeConfiguration & conf);

  void halt();
  BT::NodeStatus tick();

  static BT::PortsList providedPorts()
  {
    return BT::PortsList({});
  }
```

br2_bt_bumpgo/include/br2_bt_bumpgo/Turn.hpp

```
private:
  rclcpp::Node::SharedPtr node_;
  rclcpp::Time start_time_;
  rclcpp::Publisher<geometry_msgs::msg::Twist>::SharedPtr vel_pub_;
};

}  // namespace br2_bt_bumpgo

#endif  // BR2_BT_BUMPGO__TURN_HPP_
```

br2_bt_bumpgo/include/br2_bt_bumpgo/IsObstacle.hpp

```
#ifndef BR2_BT_BUMPGO__ISOBSTACLE_HPP_
#define BR2_BT_BUMPGO__ISOBSTACLE_HPP_

#include <string>

#include "behaviortree_cpp_v3/behavior_tree.h"
#include "behaviortree_cpp_v3/bt_factory.h"

#include "sensor_msgs/msg/laser_scan.hpp"
#include "rclcpp/rclcpp.hpp"

namespace br2_bt_bumpgo
{

class IsObstacle : public BT::ConditionNode
{
public:
  explicit IsObstacle(
    const std::string & xml_tag_name,
    const BT::NodeConfiguration & conf);

  BT::NodeStatus tick();

  static BT::PortsList providedPorts()
  {
    return BT::PortsList(
      {
        BT::InputPort<double>("distance")
      });
  }

  void laser_callback(sensor_msgs::msg::LaserScan::UniquePtr msg);

private:
  rclcpp::Node::SharedPtr node_;
  rclcpp::Time last_reading_time_;
  rclcpp::Subscription<sensor_msgs::msg::LaserScan>::SharedPtr laser_sub_;
  sensor_msgs::msg::LaserScan::UniquePtr last_scan_;
};

}  // namespace br2_bt_bumpgo

#endif  // BR2_BT_BUMPGO__ISOBSTACLE_HPP_
```

br2_bt_bumpgo/include/br2_bt_bumpgo/Back.hpp

```
#ifndef BR2_BT_BUMPGO__BACK_HPP_
#define BR2_BT_BUMPGO__BACK_HPP_

#include <string>

#include "behaviortree_cpp_v3/behavior_tree.h"
#include "behaviortree_cpp_v3/bt_factory.h"

#include "geometry_msgs/msg/twist.hpp"
#include "rclcpp/rclcpp.hpp"

namespace br2_bt_bumpgo
{

class Back : public BT::ActionNodeBase
{
public:
  explicit Back(
    const std::string & xml_tag_name,
    const BT::NodeConfiguration & conf);

  void halt();
  BT::NodeStatus tick();
```

br2_bt_bumpgo/include/br2_bt_bumpgo/Back.hpp

```
  static BT::PortsList providedPorts()
  {
    return BT::PortsList({});
  }

private:
  rclcpp::Node::SharedPtr node_;
  rclcpp::Time start_time_;
  rclcpp::Publisher<geometry_msgs::msg::Twist>::SharedPtr vel_pub_;
};

}  // namespace br2_bt_bumpgo

#endif  // BR2_BT_BUMPGO__BACK_HPP_
```

br2_bt_bumpgo/include/br2_bt_bumpgo/Forward.hpp

```
#ifndef BR2_BT_BUMPGO__FORWARD_HPP_
#define BR2_BT_BUMPGO__FORWARD_HPP_

#include <string>

#include "behaviortree_cpp_v3/behavior_tree.h"
#include "behaviortree_cpp_v3/bt_factory.h"

#include "geometry_msgs/msg/twist.hpp"
#include "rclcpp/rclcpp.hpp"

namespace br2_bt_bumpgo
{

class Forward : public BT::ActionNodeBase
{
public:
  explicit Forward(
    const std::string & xml_tag_name,
    const BT::NodeConfiguration & conf);

  void halt() {}
  BT::NodeStatus tick();
```

br2_bt_bumpgo/include/br2_bt_bumpgo/Forward.hpp

```
  static BT::PortsList providedPorts()
  {
    return BT::PortsList({});
  }

private:
  rclcpp::Node::SharedPtr node_;
  rclcpp::Publisher<geometry_msgs::msg::Twist>::SharedPtr vel_pub_;
};

}  // namespace br2_bt_bumpgo

#endif  // BR2_BT_BUMPGO__FORWARD_HPP_
```

br2_bt_bumpgo/src/bt_bumpgo_main.cpp

```
#include <string>
#include <memory>

#include "behaviortree_cpp_v3/behavior_tree.h"
#include "behaviortree_cpp_v3/bt_factory.h"
#include "behaviortree_cpp_v3/utils/shared_library.h"
#include "behaviortree_cpp_v3/loggers/bt_zmq_publisher.h"

#include "ament_index_cpp/get_package_share_directory.hpp"

#include "rclcpp/rclcpp.hpp"


int main(int argc, char * argv[])
{
  rclcpp::init(argc, argv);

  auto node = rclcpp::Node::make_shared("patrolling_node");

  BT::BehaviorTreeFactory factory;
  BT::SharedLibrary loader;

  factory.registerFromPlugin(loader.getOSName("br2_forward_bt_node"));
  factory.registerFromPlugin(loader.getOSName("br2_back_bt_node"));
  factory.registerFromPlugin(loader.getOSName("br2_turn_bt_node"));
  factory.registerFromPlugin(loader.getOSName("br2_is_obstacle_bt_node"));

  std::string pkgpath = ament_index_cpp::get_package_share_directory("br2_bt_bumpgo");
  std::string xml_file = pkgpath + "/behavior_tree_xml/bumpgo.xml";

  auto blackboard = BT::Blackboard::create();
  blackboard->set("node", node);
  BT::Tree tree = factory.createTreeFromFile(xml_file, blackboard);

  auto publisher_zmq = std::make_shared<BT::PublisherZMQ>(tree, 10, 1666, 1667);

  rclcpp::Rate rate(10);

  bool finish = false;
  while (!finish && rclcpp::ok()) {
    finish = tree.rootNode()->executeTick() != BT::NodeStatus::RUNNING;

    rclcpp::spin_some(node);
    rate.sleep();
  }

  rclcpp::shutdown();
  return 0;
}
```

br2_bt_bumpgo/src/br2_bt_bumpgo/Back.cpp

```
#include <string>
#include <iostream>

#include "br2_bt_bumpgo/Back.hpp"

#include "behaviortree_cpp_v3/behavior_tree.h"

#include "geometry_msgs/msg/twist.hpp"
#include "rclcpp/rclcpp.hpp"

namespace br2_bt_bumpgo
{

using namespace std::chrono_literals;

Back::Back(
  const std::string & xml_tag_name,
  const BT::NodeConfiguration & conf)
: BT::ActionNodeBase(xml_tag_name, conf)
{
  config().blackboard->get("node", node_);

  vel_pub_ = node_->create_publisher<geometry_msgs::msg::Twist>("/output_vel", 100);
}

void
Back::halt()
{
}

BT::NodeStatus
Back::tick()
{
  if (status() == BT::NodeStatus::IDLE) {
    start_time_ = node_->now();
  }

  geometry_msgs::msg::Twist vel_msgs;
  vel_msgs.linear.x = -0.3;
  vel_pub_->publish(vel_msgs);

  auto elapsed = node_->now() - start_time_;

  if (elapsed < 3s) {
    return BT::NodeStatus::RUNNING;
  } else {
    return BT::NodeStatus::SUCCESS;
  }
}

}  // namespace br2_bt_bumpgo

#include "behaviortree_cpp_v3/bt_factory.h"
BT_REGISTER_NODES(factory)
{
  factory.registerNodeType<br2_bt_bumpgo::Back>("Back");
}
```

br2_bt_bumpgo/src/br2_bt_bumpgo/Forward.cpp

```
#include <string>
#include <iostream>

#include "br2_bt_bumpgo/Forward.hpp"

#include "behaviortree_cpp_v3/behavior_tree.h"

#include "geometry_msgs/msg/twist.hpp"
#include "rclcpp/rclcpp.hpp"

namespace br2_bt_bumpgo
{

using namespace std::chrono_literals;

Forward::Forward(
  const std::string & xml_tag_name,
  const BT::NodeConfiguration & conf)
: BT::ActionNodeBase(xml_tag_name, conf)
{
  config().blackboard->get("node", node_);

  vel_pub_ = node_->create_publisher<geometry_msgs::msg::Twist>("/output_vel", 100);
}

BT::NodeStatus
Forward::tick()
{
  geometry_msgs::msg::Twist vel_msgs;
  vel_msgs.linear.x = 0.3;
  vel_pub_->publish(vel_msgs);

  return BT::NodeStatus::RUNNING;
}

}  // namespace br2_bt_bumpgo

#include "behaviortree_cpp_v3/bt_factory.h"
BT_REGISTER_NODES(factory)
{
  factory.registerNodeType<br2_bt_bumpgo::Forward>("Forward");
}
```

br2_bt_bumpgo/src/br2_bt_bumpgo/IsObstacle.cpp

```
#include <string>
#include <utility>

#include "br2_bt_bumpgo/IsObstacle.hpp"

#include "behaviortree_cpp_v3/behavior_tree.h"

#include "sensor_msgs/msg/laser_scan.hpp"
#include "rclcpp/rclcpp.hpp"

namespace br2_bt_bumpgo
{

using namespace std::chrono_literals;
using namespace std::placeholders;

IsObstacle::IsObstacle(
  const std::string & xml_tag_name,
  const BT::NodeConfiguration & conf)
: BT::ConditionNode(xml_tag_name, conf)
{
```

br2_bt_bumpgo/src/br2_bt_bumpgo/IsObstacle.cpp

```
  config().blackboard->get("node", node_);

  laser_sub_ = node_->create_subscription<sensor_msgs::msg::LaserScan>(
    "/input_scan", 100, std::bind(&IsObstacle::laser_callback, this, _1));

  last_reading_time_ = node_->now();
}

void
IsObstacle::laser_callback(sensor_msgs::msg::LaserScan::UniquePtr msg)
{
  last_scan_ = std::move(msg);
}

BT::NodeStatus
IsObstacle::tick()
{
  if (last_scan_ == nullptr) {
    return BT::NodeStatus::FAILURE;
  }

  double distance = 1.0;
  getInput("distance", distance);

  if (last_scan_->ranges[last_scan_->ranges.size() / 2] < distance) {
    return BT::NodeStatus::SUCCESS;
  } else {
    return BT::NodeStatus::FAILURE;
  }
}

}  // namespace br2_bt_bumpgo

#include "behaviortree_cpp_v3/bt_factory.h"
BT_REGISTER_NODES(factory)
{
  factory.registerNodeType<br2_bt_bumpgo::IsObstacle>("IsObstacle");
}
```

br2_bt_bumpgo/src/br2_bt_bumpgo/Turn.cpp

```
#include <string>
#include <iostream>

#include "br2_bt_bumpgo/Turn.hpp"

#include "behaviortree_cpp_v3/behavior_tree.h"

#include "geometry_msgs/msg/twist.hpp"
#include "rclcpp/rclcpp.hpp"

namespace br2_bt_bumpgo
{

using namespace std::chrono_literals;

Turn::Turn(
  const std::string & xml_tag_name,
  const BT::NodeConfiguration & conf)
: BT::ActionNodeBase(xml_tag_name, conf)
{
  config().blackboard->get("node", node_);

  vel_pub_ = node_->create_publisher<geometry_msgs::msg::Twist>("/output_vel", 100);
}

void
Turn::halt()
{
}
```

br2_bt_bumpgo/src/br2_bt_bumpgo/Turn.cpp

```
BT::NodeStatus
Turn::tick()
{
  if (status() == BT::NodeStatus::IDLE) {
    start_time_ = node_->now();
  }

  geometry_msgs::msg::Twist vel_msgs;
  vel_msgs.angular.z = 0.5;
  vel_pub_->publish(vel_msgs);

  auto elapsed = node_->now() - start_time_;

  if (elapsed < 3s) {
    return BT::NodeStatus::RUNNING;
  } else {
    return BT::NodeStatus::SUCCESS;
  }
}

}  // namespace br2_bt_bumpgo

#include "behaviortree_cpp_v3/bt_factory.h"
BT_REGISTER_NODES(factory)
{
  factory.registerNodeType<br2_bt_bumpgo::Turn>("Turn");
}
```

br2_bt_bumpgo/tests/bt_action_test.cpp

```
#include <string>
#include <list>
#include <memory>
#include <vector>
#include <set>

#include "behaviortree_cpp_v3/behavior_tree.h"
#include "behaviortree_cpp_v3/bt_factory.h"
#include "behaviortree_cpp_v3/utils/shared_library.h"

#include "ament_index_cpp/get_package_share_directory.hpp"

#include "geometry_msgs/msg/twist.hpp"
#include "sensor_msgs/msg/laser_scan.hpp"

#include "rclcpp/rclcpp.hpp"
#include "rclcpp_action/rclcpp_action.hpp"

#include "gtest/gtest.h"

using namespace std::placeholders;
using namespace std::chrono_literals;


class VelocitySinkNode : public rclcpp::Node
{
public:
  VelocitySinkNode()
  : Node("VelocitySink")
  {
    vel_sub_ = create_subscription<geometry_msgs::msg::Twist>(
      "/output_vel", 100, std::bind(&VelocitySinkNode::vel_callback, this, _1));
  }

  void vel_callback(geometry_msgs::msg::Twist::SharedPtr msg)
  {
    vel_msgs_.push_back(*msg);
  }
```

br2_bt_bumpgo/tests/bt_action_test.cpp

```
  std::list<geometry_msgs::msg::Twist> vel_msgs_;

private:
  rclcpp::Subscription<geometry_msgs::msg::Twist>::SharedPtr vel_sub_;
};


TEST(bt_action, turn_btn)
{
  auto node = rclcpp::Node::make_shared("turn_btn_node");
  auto node_sink = std::make_shared<VelocitySinkNode>();

  BT::BehaviorTreeFactory factory;
  BT::SharedLibrary loader;

  factory.registerFromPlugin(loader.getOSName("br2_turn_bt_node"));

  std::string xml_bt =
    R"(
    <root main_tree_to_execute = "MainTree" >
      <BehaviorTree ID="MainTree">
          <Turn />
      </BehaviorTree>
    </root>)";

  auto blackboard = BT::Blackboard::create();
  blackboard->set("node", node);
  BT::Tree tree = factory.createTreeFromText(xml_bt, blackboard);

  rclcpp::Rate rate(10);
  bool finish = false;
  while (!finish && rclcpp::ok()) {
    finish = tree.rootNode()->executeTick() == BT::NodeStatus::SUCCESS;
    rclcpp::spin_some(node_sink);
    rate.sleep();
  }

  ASSERT_FALSE(node_sink->vel_msgs_.empty());
  ASSERT_NEAR(node_sink->vel_msgs_.size(), 30, 1);

  geometry_msgs::msg::Twist & one_twist = node_sink->vel_msgs_.front();

  ASSERT_GT(one_twist.angular.z, 0.1);
  ASSERT_NEAR(one_twist.linear.x, 0.0, 0.0000001);
}

TEST(bt_action, back_btn)
{
  auto node = rclcpp::Node::make_shared("back_btn_node");
  auto node_sink = std::make_shared<VelocitySinkNode>();

  BT::BehaviorTreeFactory factory;
  BT::SharedLibrary loader;

  factory.registerFromPlugin(loader.getOSName("br2_back_bt_node"));

  std::string xml_bt =gte_node
    R"(
    <root main_tree_to_execute = "MainTree" >
      <BehaviorTree ID="MainTree">
          <Back />
      </BehaviorTree>
    </root>)";

  auto blackboard = BT::Blackboard::create();
  blackboard->set("node", node);
  BT::Tree tree = factory.createTreeFromText(xml_bt, blackboard);

  rclcpp::Rate rate(10);
  bool finish = false;
  while (!finish && rclcpp::ok()) {
    finish = tree.rootNode()->executeTick() == BT::NodeStatus::SUCCESS;
    rclcpp::spin_some(node_sink);
    rate.sleep();
  }
```

br2_bt_bumpgo/tests/bt_action_test.cpp

```
  ASSERT_FALSE(node_sink->vel_msgs_.empty());
  ASSERT_NEAR(node_sink->vel_msgs_.size(), 30, 1);

  geometry_msgs::msg::Twist & one_twist = node_sink->vel_msgs_.front();

  ASSERT_LT(one_twist.linear.x, -0.1);
  ASSERT_NEAR(one_twist.angular.z, 0.0, 0.0000001);
}

TEST(bt_action, forward_btn)
{
  auto node = rclcpp::Node::make_shared("forward_btn_node");
  auto node_sink = std::make_shared<VelocitySinkNode>();

  BT::BehaviorTreeFactory factory;
  BT::SharedLibrary loader;

  factory.registerFromPlugin(loader.getOSName("br2_forward_bt_node"));

  std::string xml_bt =
    R"(
    <root main_tree_to_execute = "MainTree" >
      <BehaviorTree ID="MainTree">
          <Forward />
      </BehaviorTree>
    </root>)";

  auto blackboard = BT::Blackboard::create();
  blackboard->set("node", node);
  BT::Tree tree = factory.createTreeFromText(xml_bt, blackboard);

  rclcpp::Rate rate(10);
  auto current_status = BT::NodeStatus::FAILURE;
  int counter = 0;
  while (counter++ < 30 && rclcpp::ok()) {
    current_status = tree.rootNode()->executeTick();
    rclcpp::spin_some(node_sink);
    rate.sleep();
  }

  ASSERT_EQ(current_status, BT::NodeStatus::RUNNING);
  ASSERT_FALSE(node_sink->vel_msgs_.empty());
  ASSERT_NEAR(node_sink->vel_msgs_.size(), 30, 1);

  geometry_msgs::msg::Twist & one_twist = node_sink->vel_msgs_.front();

  ASSERT_GT(one_twist.linear.x, 0.1);
  ASSERT_NEAR(one_twist.angular.z, 0.0, 0.0000001);
}

TEST(bt_action, is_obstacle_btn)
{
  auto node = rclcpp::Node::make_shared("is_obstacle_btn_node");
  auto scan_pub = node->create_publisher<sensor_msgs::msg::LaserScan>("input_scan", 1);

  BT::BehaviorTreeFactory factory;
  BT::SharedLibrary loader;

  factory.registerFromPlugin(loader.getOSName("br2_is_obstacle_bt_node"));

  std::string xml_bt =
    R"(
    <root main_tree_to_execute = "MainTree" >
      <BehaviorTree ID="MainTree">
          <IsObstacle/>
      </BehaviorTree>
    </root>)";

  auto blackboard = BT::Blackboard::create();
  blackboard->set("node", node);
  BT::Tree tree = factory.createTreeFromText(xml_bt, blackboard);

  rclcpp::Rate rate(10);

  sensor_msgs::msg::LaserScan scan;
  scan.ranges.push_back(2.0);
  for (int i = 0; i < 10; i++) {
```

br2_bt_bumpgo/tests/bt_action_test.cpp

```
    scan_pub->publish(scan);
    rclcpp::spin_some(node);
    rate.sleep();
  }

  BT::NodeStatus current_status = tree.rootNode()->executeTick();
  ASSERT_EQ(current_status, BT::NodeStatus::FAILURE);

  scan.ranges[0] = 0.3;
  for (int i = 0; i < 10; i++) {
    scan_pub->publish(scan);
    rclcpp::spin_some(node);
    rate.sleep();
  }

  current_status = tree.rootNode()->executeTick();
  ASSERT_EQ(current_status, BT::NodeStatus::SUCCESS);

  xml_bt =
    R"(
    <root main_tree_to_execute = "MainTree" >
      <BehaviorTree ID="MainTree">
          <IsObstacle distance="0.5"/>
      </BehaviorTree>
    </root>)";
  tree = factory.createTreeFromText(xml_bt, blackboard);

  scan.ranges[0] = 0.3;
  for (int i = 0; i < 10; i++) {
    scan_pub->publish(scan);
    rclcpp::spin_some(node);
    rate.sleep();
  }

  current_status = tree.rootNode()->executeTick();
  ASSERT_EQ(current_status, BT::NodeStatus::SUCCESS);

  scan.ranges[0] = 0.6;
  for (int i = 0; i < 10; i++) {
    scan_pub->publish(scan);
    rclcpp::spin_some(node);
    rate.sleep();
  }

  current_status = tree.rootNode()->executeTick();
  ASSERT_EQ(current_status, BT::NodeStatus::FAILURE);
}

int main(int argc, char ** argv)
{
  rclcpp::init(argc, argv);

  testing::InitGoogleTest(&argc, argv);
  return RUN_ALL_TESTS();
}
```

br2_bt_bumpgo/tests/CMakeLists.txt

```
ament_add_gtest(bt_action_test bt_action_test.cpp)
ament_target_dependencies(bt_action_test ${dependencies})

add_executable(bt_forward bt_forward_main.cpp)
ament_target_dependencies(bt_forward ${dependencies})
target_link_libraries(bt_forward ${ZMQ_LIBRARIES})

install(TARGETS
  bt_forward
  ARCHIVE DESTINATION lib
  LIBRARY DESTINATION lib
  RUNTIME DESTINATION lib/${PROJECT_NAME}
)
```

br2_bt_bumpgo/tests/bt_forward_main.cpp

```
#include <string>
#include <memory>

#include "behaviortree_cpp_v3/behavior_tree.h"
#include "behaviortree_cpp_v3/bt_factory.h"
#include "behaviortree_cpp_v3/utils/shared_library.h"
#include "behaviortree_cpp_v3/loggers/bt_zmq_publisher.h"

#include "ament_index_cpp/get_package_share_directory.hpp"

#include "rclcpp/rclcpp.hpp"


int main(int argc, char * argv[])
{
  rclcpp::init(argc, argv);

  auto node = rclcpp::Node::make_shared("forward_node");

  BT::BehaviorTreeFactory factory;
  BT::SharedLibrary loader;

  factory.registerFromPlugin(loader.getOSName("br2_forward_bt_node"));

  std::string xml_bt =
    R"(
    <root main_tree_to_execute = "MainTree" >
      <BehaviorTree ID="MainTree">
          <Forward />
      </BehaviorTree>
    </root>)";

  auto blackboard = BT::Blackboard::create();
  blackboard->set("node", node);
  BT::Tree tree = factory.createTreeFromText(xml_bt, blackboard);

  rclcpp::Rate rate(10);
  bool finish = false;
  while (!finish && rclcpp::ok()) {
    finish = tree.rootNode()->executeTick() != BT::NodeStatus::RUNNING;

    rclcpp::spin_some(node);
    rate.sleep();
  }

  rclcpp::shutdown();
  return 0;
}
```

A.9 PACKAGE BR2_BT_PATROLLING

Package br2_bt_patrolling

```
br2_bt_patrolling
├── behavior_tree_xml
│   └── patrolling.xml
├── cmake
│   └── FindZMQ.cmake
├── CMakeLists.txt
├── include
│   └── br2_bt_patrolling
│       ├── BatteryChecker.hpp
│       ├── ctrl_support
│       │   ├── BTActionNode.hpp
│       │   └── BTLifecycleCtrlNode.hpp
│       ├── GetWaypoint.hpp
│       ├── Move.hpp
│       ├── Patrol.hpp
│       ├── Recharge.hpp
│       └── TrackObjects.hpp
├── launch
│   └── patrolling.launch.py
├── package.xml
├── src
│   ├── br2_bt_patrolling
│   │   ├── BatteryChecker.cpp
│   │   ├── GetWaypoint.cpp
│   │   ├── Move.cpp
│   │   ├── Patrol.cpp
│   │   ├── Recharge.cpp
│   │   └── TrackObjects.cpp
│   └── patrolling_main.cpp
└── tests
    ├── bt_action_test.cpp
    └── CMakeLists.txt
```

br2_bt_patrolling/behavior_tree_xml/patrolling.xml

```xml
<?xml version="1.0"?>
<root main_tree_to_execute="BehaviorTree">
    <!-- ////////// -->
    <BehaviorTree ID="BehaviorTree">
        <KeepRunningUntilFailure>
            <ReactiveSequence>
                <Fallback>
                    <Action ID="BatteryChecker"/>
                    <Sequence>
                        <Action ID="GetWaypoint" waypoint="{recharge_wp}" wp_id="recharge"/>
                        <Action ID="Move" goal="{recharge_wp}"/>
                        <Action ID="Recharge"/>
                    </Sequence>
                </Fallback>
                <Sequence>
                    <Action ID="GetWaypoint" waypoint="{wp}" wp_id="next"/>
                        <Parallel success_threshold="1" failure_threshold="1">
                          <Action ID="TrackObjects"/>
                          <Action ID="Move" goal="{wp}"/>
                        </Parallel>
                    <Action ID="Patrol"/>
                </Sequence>
            </ReactiveSequence>
        </KeepRunningUntilFailure>
    </BehaviorTree>
```

br2_bt_patrolling/behavior_tree_xml/patrolling.xml

```
    <!-- ////////// -->
    <TreeNodesModel>
        <Action ID="BatteryChecker"/>
        <Action ID="GetWaypoint">
            <output_port name="waypoint"/>
            <input_port name="wp_id"/>
        </Action>
        <Action ID="Move">
            <input_port name="goal"/>
        </Action>
        <Action ID="Patrol"/>
        <Action ID="Recharge"/>
        <Action ID="TrackObjects"/>
    </TreeNodesModel>
    <!-- ////////// -->
</root>
```

br2_bt_patrolling/CMakeLists.txt

```
cmake_minimum_required(VERSION 3.5)
project(br2_bt_patrolling)

set(CMAKE_BUILD_TYPE Debug)

set(CMAKE_CONFIG_PATH ${CMAKE_MODULE_PATH}  "${CMAKE_CURRENT_LIST_DIR}/cmake")
list(APPEND CMAKE_MODULE_PATH "${CMAKE_CONFIG_PATH}")

find_package(ament_cmake REQUIRED)
find_package(rclcpp REQUIRED)
find_package(rclcpp_lifecycle REQUIRED)
find_package(rclcpp_action REQUIRED)
find_package(behaviortree_cpp_v3 REQUIRED)
find_package(action_msgs REQUIRED)
find_package(lifecycle_msgs REQUIRED)
find_package(geometry_msgs REQUIRED)
find_package(nav2_msgs REQUIRED)
find_package(ament_index_cpp REQUIRED)

find_package(ZMQ)
if(ZMQ_FOUND)
    message(STATUS "ZeroMQ found.")
    add_definitions(-DZMQ_FOUND)
else()
  message(WARNING "ZeroMQ NOT found. Not including PublisherZMQ.")
endif()

set(CMAKE_CXX_STANDARD 17)

set(dependencies
    rclcpp
    rclcpp_lifecycle
    rclcpp_action
    behaviortree_cpp_v3
    action_msgs
    lifecycle_msgs
    geometry_msgs
    nav2_msgs
    ament_index_cpp
)

include_directories(include ${ZMQ_INCLUDE_DIRS})
```

br2_bt_patrolling/CMakeLists.txt

```
add_library(br2_recharge_bt_node SHARED src/br2_bt_patrolling/Recharge.cpp)
add_library(br2_patrol_bt_node SHARED src/br2_bt_patrolling/Patrol.cpp)
add_library(br2_move_bt_node SHARED src/br2_bt_patrolling/Move.cpp)
add_library(br2_get_waypoint_bt_node SHARED src/br2_bt_patrolling/GetWaypoint.cpp)
add_library(br2_battery_checker_bt_node SHARED src/br2_bt_patrolling/BatteryChecker.cpp)
add_library(br2_track_objects_bt_node SHARED src/br2_bt_patrolling/TrackObjects.cpp)
list(APPEND plugin_libs
  br2_recharge_bt_node
  br2_patrol_bt_node
  br2_move_bt_node
  br2_get_waypoint_bt_node
  br2_battery_checker_bt_node
  br2_track_objects_bt_node
)

foreach(bt_plugin ${plugin_libs})
  ament_target_dependencies(${bt_plugin} ${dependencies})
  target_compile_definitions(${bt_plugin} PRIVATE BT_PLUGIN_EXPORT)
endforeach()

add_executable(patrolling_main src/patrolling_main.cpp)
ament_target_dependencies(patrolling_main ${dependencies})
target_link_libraries(patrolling_main ${ZMQ_LIBRARIES})

install(TARGETS
  ${plugin_libs}
  patrolling_main
  ARCHIVE DESTINATION lib
  LIBRARY DESTINATION lib
  RUNTIME DESTINATION lib/${PROJECT_NAME}
)

install(DIRECTORY include/
  DESTINATION include/
)

install(DIRECTORY behavior_tree_xml launch
  DESTINATION share/${PROJECT_NAME}
)


if(BUILD_TESTING)
  find_package(ament_lint_auto REQUIRED)
  ament_lint_auto_find_test_dependencies()

  set(ament_cmake_cpplint_FOUND TRUE)
  ament_lint_auto_find_test_dependencies()

  find_package(ament_cmake_gtest REQUIRED)

  add_subdirectory(tests)
endif()

ament_export_include_directories(include)
ament_export_dependencies(${dependencies})

ament_package()
```

br2_bt_patrolling/launch/patrolling.launch.py

```
import os

from ament_index_python.packages import get_package_share_directory

from launch import LaunchDescription
from launch.actions import IncludeLaunchDescription
from launch.launch_description_sources import PythonLaunchDescriptionSource
from launch_ros.actions import Node


def generate_launch_description():

    tracking_dir = get_package_share_directory('br2_tracking')

    tracking_cmd = IncludeLaunchDescription(
        PythonLaunchDescriptionSource(
          os.path.join(tracking_dir, 'launch', 'tracking.launch.py')))

    patrolling_cmd = Node(
        package='br2_bt_patrolling',
        executable='patrolling_main',
        parameters=[{
          'use_sim_time': True
        }],
        remappings=[
          ('input_scan', '/scan_raw'),
          ('output_vel', '/nav_vel')
        ],
        output='screen'
    )

    ld = LaunchDescription()

    # Add any actions
    ld.add_action(tracking_cmd)
    ld.add_action(patrolling_cmd)

    return ld
```

br2_bt_patrolling/package.xml

```
<?xml version="1.0"?>
<?xml-model href="http://download.ros.org/schema/package_format3.xsd"
  schematypens="http://www.w3.org/2001/XMLSchema"?>
<package format="3">
  <name>br2_bt_patrolling</name>
  <version>0.0.0</version>
  <description>Patrolling behavior package</description>
  <maintainer email="fmrico@gmail.com">Francisco Martín</maintainer>
  <license>Apache 2.0</license>

  <buildtool_depend>ament_cmake</buildtool_depend>

  <depend>rclcpp</depend>
  <depend>rclcpp_lifecycle</depend>
  <depend>rclcpp_action</depend>
  <depend>behaviortree_cpp_v3</depend>
  <depend>action_msgs</depend>
  <depend>geometry_msgs</depend>
  <depend>lifecycle_msgs</depend>
  <depend>nav2_msgs</depend>
  <depend>libzmq3-dev</depend>
  <depend>ament_index_cpp</depend>

  <test_depend>ament_lint_auto</test_depend>
  <test_depend>ament_lint_common</test_depend>
  <test_depend>ament_cmake_gtest</test_depend>

  <export>
    <build_type>ament_cmake</build_type>
  </export>
</package>
```

br2_bt_patrolling/include/br2_bt_patrolling/BatteryChecker.hpp

```
#ifndef BR2_BT_PATROLLING__BATTERYCHECKER_HPP_
#define BR2_BT_PATROLLING__BATTERYCHECKER_HPP_

#include <string>
#include <vector>

#include "behaviortree_cpp_v3/behavior_tree.h"
#include "behaviortree_cpp_v3/bt_factory.h"

#include "geometry_msgs/msg/twist.hpp"

#include "rclcpp/rclcpp.hpp"

namespace br2_bt_patrolling
{

class BatteryChecker : public BT::ConditionNode
{
public:
  explicit BatteryChecker(
    const std::string & xml_tag_name,
    const BT::NodeConfiguration & conf);

  BT::NodeStatus tick();

  static BT::PortsList providedPorts()
  {
    return BT::PortsList({});
  }

  void vel_callback(const geometry_msgs::msg::Twist::SharedPtr msg);

  const float DECAY_LEVEL = 0.5;  // 0.5 * |vel| * dt
  const float EPSILON = 0.01;  // 0.001 * dt
  const float MIN_LEVEL = 10.0;

private:
  void update_battery();

  rclcpp::Node::SharedPtr node_;
  rclcpp::Time last_reading_time_;
  geometry_msgs::msg::Twist last_twist_;
  rclcpp::Subscription<geometry_msgs::msg::Twist>::SharedPtr vel_sub_;
};

}  // namespace br2_bt_patrolling

#endif  // BR2_BT_PATROLLING__BATTERYCHECKER_HPP_
```

br2_bt_patrolling/include/br2_bt_patrolling/ctrl_support/BTLifecycleCtrlNode.hpp

```
#ifndef BR2_BT_PATROLLING__CTRL_SUPPORT__BTLIFECYCLECTRLNODE_HPP_
#define BR2_BT_PATROLLING__CTRL_SUPPORT__BTLIFECYCLECTRLNODE_HPP_

#include <memory>
#include <string>

#include "lifecycle_msgs/srv/change_state.hpp"
#include "lifecycle_msgs/srv/get_state.hpp"
#include "lifecycle_msgs/msg/state.hpp"

#include "behaviortree_cpp_v3/action_node.h"
#include "rclcpp/rclcpp.hpp"
namespace br2_bt_patrolling

{

using namespace std::chrono_literals;  // NOLINT
```

br2_bt_patrolling/include/br2_bt_patrolling/ctrl_support/BTLifecycleCtrlNode.hpp

```
class BtLifecycleCtrlNode : public BT::ActionNodeBase
{
public:
  BtLifecycleCtrlNode(
    const std::string & xml_tag_name,
    const std::string & node_name,
    const BT::NodeConfiguration & conf)
  : BT::ActionNodeBase(xml_tag_name, conf), ctrl_node_name_(node_name)
  {
    node_ = config().blackboard->get<rclcpp::Node::SharedPtr>("node");
  }

  BtLifecycleCtrlNode() = delete;

  virtual ~BtLifecycleCtrlNode()
  {
  }

  template<typename serviceT>
  typename rclcpp::Client<serviceT>::SharedPtr createServiceClient(
    const std::string & service_name)
  {
    auto srv = node_->create_client<serviceT>(service_name);
    while (!srv->wait_for_service(1s)) {
      if (!rclcpp::ok()) {
        RCLCPP_ERROR(node_->get_logger(), "Interrupted while waiting for the service.");
      } else {
        RCLCPP_INFO(node_->get_logger(), "service not available, waiting again...");
      }
    }
    return srv;
  }

  virtual void on_tick() {}

  virtual BT::NodeStatus on_success()
  {
    return BT::NodeStatus::SUCCESS;
  }

  virtual BT::NodeStatus on_failure()
  {
    return BT::NodeStatus::FAILURE;
  }

  BT::NodeStatus tick() override
  {
    if (status() == BT::NodeStatus::IDLE) {
      change_state_client_ = createServiceClient<lifecycle_msgs::srv::ChangeState>(
        ctrl_node_name_ + "/change_state");
      get_state_client_ = createServiceClient<lifecycle_msgs::srv::GetState>(
        ctrl_node_name_ + "/get_state");
    }

    if (ctrl_node_state_ != lifecycle_msgs::msg::State::PRIMARY_STATE_ACTIVE) {
      ctrl_node_state_ = get_state();
      set_state(lifecycle_msgs::msg::State::PRIMARY_STATE_ACTIVE);
    }

    on_tick();

    return BT::NodeStatus::RUNNING;
  }

  void halt() override
  {
    if (ctrl_node_state_ == lifecycle_msgs::msg::State::PRIMARY_STATE_ACTIVE) {
      set_state(lifecycle_msgs::msg::State::PRIMARY_STATE_INACTIVE);
    }
    setStatus(BT::NodeStatus::IDLE);
  }

  // Get the state of the controlled node
  uint8_t get_state()
```

br2_bt_patrolling/include/br2_bt_patrolling/ctrl_support/BTLifecycleCtrlNode.hpp

```
  {
    auto request = std::make_shared<lifecycle_msgs::srv::GetState::Request>();
    auto result = get_state_client_->async_send_request(request);

    if (rclcpp::spin_until_future_complete(node_, result) !=
      rclcpp::FutureReturnCode::SUCCESS)
    {
      lifecycle_msgs::msg::State get_state;

      RCLCPP_ERROR(node_->get_logger(), "Failed to call get_state service");
      return lifecycle_msgs::msg::State::PRIMARY_STATE_UNKNOWN;
    }

    return result.get()->current_state.id;
  }

  // Get the state of the controlled node. Ot can fail, if not transition possible
  bool set_state(uint8_t state)
  {
    auto request = std::make_shared<lifecycle_msgs::srv::ChangeState::Request>();

    if (state == lifecycle_msgs::msg::State::PRIMARY_STATE_ACTIVE &&
      ctrl_node_state_ == lifecycle_msgs::msg::State::PRIMARY_STATE_INACTIVE)
    {
      request->transition.id = lifecycle_msgs::msg::Transition::TRANSITION_ACTIVATE;
    } else {
      if (state == lifecycle_msgs::msg::State::PRIMARY_STATE_INACTIVE &&
        ctrl_node_state_ == lifecycle_msgs::msg::State::PRIMARY_STATE_ACTIVE)
      {
        request->transition.id = lifecycle_msgs::msg::Transition::TRANSITION_DEACTIVATE;
      } else {
        if (state != ctrl_node_state_) {
          RCLCPP_ERROR(
            node_->get_logger(),
            "Transition not possible %zu -> %zu", ctrl_node_state_, state);
          return false;
        } else {
          return true;
        }
      }
    }

    auto result = change_state_client_->async_send_request(request);

    if (rclcpp::spin_until_future_complete(node_, result) !=
      rclcpp::FutureReturnCode::SUCCESS)
    {
      RCLCPP_ERROR(node_->get_logger(), "Failed to call set_state service");
      return false;
    }

    if (!result.get()->success) {
      RCLCPP_ERROR(
        node_->get_logger(),
        "Failed to set node state %zu -> %zu", ctrl_node_state_, state);
      return false;
    } else {
      RCLCPP_INFO(
        node_->get_logger(), "Transition success  %zu -> %zu", ctrl_node_state_, state);
    }

    ctrl_node_state_ = state;
    return true;
  }

  std::string ctrl_node_name_;
  uint8_t ctrl_node_state_;

  rclcpp::Client<lifecycle_msgs::srv::ChangeState>::SharedPtr change_state_client_;
  rclcpp::Client<lifecycle_msgs::srv::GetState>::SharedPtr get_state_client_;

  rclcpp::Node::SharedPtr node_;
};


}  // namespace br2_bt_patrolling

#endif  // BR2_BT_PATROLLING__CTRL_SUPPORT__BTLIFECYCLECTRLNODE_HPP_
```

br2_bt_patrolling/include/br2_bt_patrolling/ctrl_support/BTActionNode.hpp

```
// Copyright (c) 2018 Intel Corporation
//
// Licensed under the Apache License, Version 2.0 (the "License");
// you may not use this file except in compliance with the License.
// You may obtain a copy of the License at
//
//     http://www.apache.org/licenses/LICENSE-2.0
//
// Unless required by applicable law or agreed to in writing, software
// distributed under the License is distributed on an "AS IS" BASIS,
// WITHOUT WARRANTIES OR CONDITIONS OF ANY KIND, either express or implied.
// See the License for the specific language governing permissions and
// limitations under the License.

#ifndef BR2_BT_PATROLLING__CTRL_SUPPORT__BTACTIONNODE_HPP_
#define BR2_BT_PATROLLING__CTRL_SUPPORT__BTACTIONNODE_HPP_

#include <memory>
#include <string>

#include "behaviortree_cpp_v3/action_node.h"
#include "rclcpp/rclcpp.hpp"
#include "rclcpp_action/rclcpp_action.hpp"

namespace br2_bt_patrolling
{

using namespace std::chrono_literals;  // NOLINT

template<class ActionT, class NodeT = rclcpp::Node>
class BtActionNode : public BT::ActionNodeBase
{
public:
  BtActionNode(
    const std::string & xml_tag_name,
    const std::string & action_name,
    const BT::NodeConfiguration & conf)
  : BT::ActionNodeBase(xml_tag_name, conf), action_name_(action_name)
  {
    node_ = config().blackboard->get<typename NodeT::SharedPtr>("node");

    server_timeout_ = 1s;

    // Initialize the input and output messages
    goal_ = typename ActionT::Goal();
    result_ = typename rclcpp_action::ClientGoalHandle<ActionT>::WrappedResult();

    std::string remapped_action_name;
    if (getInput("server_name", remapped_action_name)) {
      action_name_ = remapped_action_name;
    }
    createActionClient(action_name_);

    // Give the derive class a chance to do any initialization
    RCLCPP_INFO(
      node_->get_logger(), "\"%s\" BtActionNode initialized", xml_tag_name.c_str());
  }

  BtActionNode() = delete;

  virtual ~BtActionNode()
  {
  }

  // Create instance of an action server
  void createActionClient(const std::string & action_name)
  {
    // Now that we have the ROS node to use, create the action client for this BT action
    action_client_ = rclcpp_action::create_client<ActionT>(node_, action_name);

    // Make sure the server is actually there before continuing
    RCLCPP_INFO(
      node_->get_logger(), "Waiting for \"%s\" action server", action_name.c_str());
    action_client_->wait_for_action_server();
  }
```

br2_bt_patrolling/include/br2_bt_patrolling/ctrl_support/BTActionNode.hpp

```
// Any subclass of BtActionNode that accepts parameters must provide a
// providedPorts method and call providedBasicPorts in it.
static BT::PortsList providedBasicPorts(BT::PortsList addition)
{
  BT::PortsList basic = {
    BT::InputPort<std::string>("server_name", "Action server name"),
    BT::InputPort<std::chrono::milliseconds>("server_timeout")
  };
  basic.insert(addition.begin(), addition.end());

  return basic;
}

static BT::PortsList providedPorts()
{
  return providedBasicPorts({});
}

// Derived classes can override any of the following methods to hook into the
// processing for the action: on_tick, on_wait_for_result, and on_success

// Could do dynamic checks, such as getting updates to values on the blackboard
virtual void on_tick()
{
}

// There can be many loop iterations per tick. Any opportunity to do something after
// a timeout waiting for a result that hasn't been received yet
virtual void on_wait_for_result()
{
}

// Called upon successful completion of the action. A derived class can override this
// method to put a value on the blackboard, for example.
virtual BT::NodeStatus on_success()
{
  return BT::NodeStatus::SUCCESS;
}

// Called when a the action is aborted. By default, the node will return FAILURE.
// The user may override it to return another value, instead.
virtual BT::NodeStatus on_aborted()
{
  return BT::NodeStatus::FAILURE;
}

// Called when a the action is cancelled. By default, the node will return SUCCESS.
// The user may override it to return another value, instead.
virtual BT::NodeStatus on_cancelled()
{
  return BT::NodeStatus::SUCCESS;
}

// The main override required by a BT action
BT::NodeStatus tick() override
{
  // first step to be done only at the beginning of the Action
  if (status() == BT::NodeStatus::IDLE) {
    createActionClient(action_name_);

    // setting the status to RUNNING to notify the BT Loggers (if any)
    setStatus(BT::NodeStatus::RUNNING);

    // user defined callback
    on_tick();

    on_new_goal_received();
  }

  // The following code corresponds to the "RUNNING" loop
  if (rclcpp::ok() && !goal_result_available_) {
   // user defined callback. May modify the value of "goal_updated_"
   on_wait_for_result();

   auto goal_status = goal_handle_->get_status();
   if (goal_updated_ && (goal_status == action_msgs::msg::GoalStatus::STATUS_EXECUTING ||
     goal_status == action_msgs::msg::GoalStatus::STATUS_ACCEPTED))
```

br2_bt_patrolling/include/br2_bt_patrolling/ctrl_support/BTActionNode.hpp

```
      {
        goal_updated_ = false;
        on_new_goal_received();
      }

      rclcpp::spin_some(node_->get_node_base_interface());

      // check if, after invoking spin_some(), we finally received the result
      if (!goal_result_available_) {
        // Yield this Action, returning RUNNING
        return BT::NodeStatus::RUNNING;
      }
    }

    switch (result_.code) {
      case rclcpp_action::ResultCode::SUCCEEDED:
        return on_success();

      case rclcpp_action::ResultCode::ABORTED:
        return on_aborted();

      case rclcpp_action::ResultCode::CANCELED:
        return on_cancelled();

      default:
        throw std::logic_error("BtActionNode::Tick: invalid status value");
    }
  }

  // The other (optional) override required by a BT action. In this case, we
  // make sure to cancel the ROS2 action if it is still running.
  void halt() override
  {
    if (should_cancel_goal()) {
      auto future_cancel = action_client_->async_cancel_goal(goal_handle_);
      if (rclcpp::spin_until_future_complete(
          node_->get_node_base_interface(), future_cancel, server_timeout_) !=
        rclcpp::FutureReturnCode::SUCCESS)
      {
        RCLCPP_ERROR(
          node_->get_logger(),
          "Failed to cancel action server for %s", action_name_.c_str());
      }
    }

    setStatus(BT::NodeStatus::IDLE);
  }

protected:
  bool should_cancel_goal()
  {
    // Shut the node down if it is currently running
    if (status() != BT::NodeStatus::RUNNING) {
      return false;
    }

    rclcpp::spin_some(node_->get_node_base_interface());
    auto status = goal_handle_->get_status();

    // Check if the goal is still executing
    return status == action_msgs::msg::GoalStatus::STATUS_ACCEPTED ||
           status == action_msgs::msg::GoalStatus::STATUS_EXECUTING;
  }


  void on_new_goal_received()
  {
    goal_result_available_ = false;
    auto send_goal_options = typename rclcpp_action::Client<ActionT>::SendGoalOptions();
    send_goal_options.result_callback =
      [this](const typename
        rclcpp_action::ClientGoalHandle<ActionT>::WrappedResult & result) {

        // TODO(#1652): a work around until rcl_action interface is updated
        // if goal ids are not matched, the older goal call this callback so ignore
        //  the result if matched, it must be processed (including aborted)
```

br2_bt_patrolling/include/br2_bt_patrolling/ctrl_support/BTActionNode.hpp

```
        if (this->goal_handle_->get_goal_id() == result.goal_id) {
          goal_result_available_ = true;
          result_ = result;
        }
      };

    auto future_goal_handle = action_client_->async_send_goal(goal_, send_goal_options);

    if (rclcpp::spin_until_future_complete(
        node_->get_node_base_interface(), future_goal_handle, server_timeout_) !=
      rclcpp::FutureReturnCode::SUCCESS)
    {
      throw std::runtime_error("send_goal failed");
    }

    goal_handle_ = future_goal_handle.get();
    if (!goal_handle_) {
      throw std::runtime_error("Goal was rejected by the action server");
    }
  }

  void increment_recovery_count()
  {
    int recovery_count = 0;
    config().blackboard->get<int>("number_recoveries", recovery_count);  // NOLINT
    recovery_count += 1;
    config().blackboard->set<int>("number_recoveries", recovery_count);  // NOLINT
  }

  std::string action_name_;
  typename std::shared_ptr<rclcpp_action::Client<ActionT>> action_client_;

  // All ROS2 actions have a goal and a result
  typename ActionT::Goal goal_;
  bool goal_updated_{false};
  bool goal_result_available_{false};
  typename rclcpp_action::ClientGoalHandle<ActionT>::SharedPtr goal_handle_;
  typename rclcpp_action::ClientGoalHandle<ActionT>::WrappedResult result_;

  // The node that will be used for any ROS operations
  typename NodeT::SharedPtr node_;

  // The timeout value while waiting for response from a server when a
  // new action goal is sent or canceled
  std::chrono::milliseconds server_timeout_;
};


}  // namespace br2_bt_patrolling

#endif  // BR2_BT_PATROLLING__CTRL_SUPPORT__BTACTIONNODE_HPP_
```

br2_bt_patrolling/include/br2_bt_patrolling/Recharge.hpp

```
#ifndef BR2_BT_PATROLLING__RECHARGE_HPP_
#define BR2_BT_PATROLLING__RECHARGE_HPP_

#include <string>

#include "behaviortree_cpp_v3/behavior_tree.h"
#include "behaviortree_cpp_v3/bt_factory.h"

namespace br2_bt_patrolling
{

class Recharge : public BT::ActionNodeBase
{
public:
  explicit Recharge(
    const std::string & xml_tag_name,
    const BT::NodeConfiguration & conf);

  void halt();
  BT::NodeStatus tick();
```

br2_bt_patrolling/include/br2_bt_patrolling/Recharge.hpp

```
  static BT::PortsList providedPorts()
  {
    return BT::PortsList({});
  }

private:
  int counter_;
};

}  // namespace br2_bt_patrolling

#endif  // BR2_BT_PATROLLING__RECHARGE_HPP_
```

br2_bt_patrolling/include/br2_bt_patrolling/GetWaypoint.hpp

```
#ifndef BR2_BT_PATROLLING__GETWAYPOINT_HPP_
#define BR2_BT_PATROLLING__GETWAYPOINT_HPP_

#include <string>
#include <vector>

#include "behaviortree_cpp_v3/behavior_tree.h"
#include "behaviortree_cpp_v3/bt_factory.h"

#include "geometry_msgs/msg/pose_stamped.hpp"

namespace br2_bt_patrolling
{

class GetWaypoint : public BT::ActionNodeBase
{
public:
  explicit GetWaypoint(
    const std::string & xml_tag_name,
    const BT::NodeConfiguration & conf);

  void halt();
  BT::NodeStatus tick();

  static BT::PortsList providedPorts()
  {
    return BT::PortsList(
      {
        BT::InputPort<std::string>("wp_id"),
        BT::OutputPort<geometry_msgs::msg::PoseStamped>("waypoint")
      });
  }

private:
  geometry_msgs::msg::PoseStamped recharge_point_;
  std::vector<geometry_msgs::msg::PoseStamped> waypoints_;
  static int current_;
};

}  // namespace br2_bt_patrolling

#endif  // BR2_BT_PATROLLING__GETWAYPOINT_HPP_
```

br2_bt_patrolling/include/br2_bt_patrolling/Patrol.hpp

```
#ifndef BR2_BT_PATROLLING__PATROL_HPP_
#define BR2_BT_PATROLLING__PATROL_HPP_

#include <string>

#include "behaviortree_cpp_v3/behavior_tree.h"
#include "behaviortree_cpp_v3/bt_factory.h"

#include "geometry_msgs/msg/twist.hpp"

#include "rclcpp/rclcpp.hpp"

namespace br2_bt_patrolling
{

class Patrol : public BT::ActionNodeBase
{
public:
  explicit Patrol(
    const std::string & xml_tag_name,
    const BT::NodeConfiguration & conf);

  void halt();
  BT::NodeStatus tick();

  static BT::PortsList providedPorts()
  {
    return BT::PortsList({});
  }

private:
  rclcpp::Node::SharedPtr node_;
  rclcpp::Time start_time_;
  rclcpp::Publisher<geometry_msgs::msg::Twist>::SharedPtr vel_pub_;
};

}  // namespace br2_bt_patrolling

#endif  // BR2_BT_PATROLLING__PATROL_HPP_
```

br2_bt_patrolling/include/br2_bt_patrolling/TrackObjects.hpp

```
#ifndef BR2_BT_PATROLLING__TRACKOBJECTS_HPP_
#define BR2_BT_PATROLLING__TRACKOBJECTS_HPP_

#include <string>

#include "geometry_msgs/msg/pose_stamped.hpp"
#include "nav2_msgs/action/navigate_to_pose.hpp"

#include "br2_bt_patrolling/ctrl_support/BTLifecycleCtrlNode.hpp"
#include "behaviortree_cpp_v3/behavior_tree.h"
#include "behaviortree_cpp_v3/bt_factory.h"

namespace br2_bt_patrolling
{

class TrackObjects : public br2_bt_patrolling::BtLifecycleCtrlNode
{
public:
  explicit TrackObjects(
    const std::string & xml_tag_name,
    const std::string & node_name,
    const BT::NodeConfiguration & conf);
```

br2_bt_patrolling/include/br2_bt_patrolling/TrackObjects.hpp

```
  static BT::PortsList providedPorts()
  {
    return BT::PortsList({});
  }
};

}  // namespace br2_bt_patrolling

#endif  // BR2_BT_PATROLLING__TRACKOBJECTS_HPP_
```

br2_bt_patrolling/include/br2_bt_patrolling/Move.hpp

```
#ifndef BR2_BT_PATROLLING__MOVE_HPP_
#define BR2_BT_PATROLLING__MOVE_HPP_

#include <string>

#include "geometry_msgs/msg/pose_stamped.hpp"
#include "nav2_msgs/action/navigate_to_pose.hpp"

#include "br2_bt_patrolling/ctrl_support/BTActionNode.hpp"
#include "behaviortree_cpp_v3/behavior_tree.h"
#include "behaviortree_cpp_v3/bt_factory.h"

namespace br2_bt_patrolling
{

class Move : public br2_bt_patrolling::BtActionNode<nav2_msgs::action::NavigateToPose>
{
public:
  explicit Move(
    const std::string & xml_tag_name,
    const std::string & action_name,
    const BT::NodeConfiguration & conf);

  void on_tick() override;
  BT::NodeStatus on_success() override;

  static BT::PortsList providedPorts()
  {
    return {
      BT::InputPort<geometry_msgs::msg::PoseStamped>("goal")
    };
  }
};

}  // namespace br2_bt_patrolling

#endif  // BR2_BT_PATROLLING__MOVE_HPP_
```

br2_bt_patrolling/src/br2_bt_patrolling/Patrol.cpp

```
#include <string>
#include <iostream>

#include "br2_bt_patrolling/Patrol.hpp"

#include "behaviortree_cpp_v3/behavior_tree.h"

#include "geometry_msgs/msg/twist.hpp"

#include "rclcpp/rclcpp.hpp"

namespace br2_bt_patrolling
{

using namespace std::chrono_literals;

Patrol::Patrol(
  const std::string & xml_tag_name,
  const BT::NodeConfiguration & conf)
: BT::ActionNodeBase(xml_tag_name, conf)
{
  config().blackboard->get("node", node_);

  vel_pub_ = node_->create_publisher<geometry_msgs::msg::Twist>("/output_vel", 100);
}

void
Patrol::halt()
{
  std::cout << "Patrol halt" << std::endl;
}

BT::NodeStatus
Patrol::tick()
{
  if (status() == BT::NodeStatus::IDLE) {
    start_time_ = node_->now();
  }

  geometry_msgs::msg::Twist vel_msgs;
  vel_msgs.angular.z = 0.5;
  vel_pub_->publish(vel_msgs);

  auto elapsed = node_->now() - start_time_;

  if (elapsed < 15s) {
    return BT::NodeStatus::RUNNING;
  } else {
    return BT::NodeStatus::SUCCESS;
  }
}

}  // namespace br2_bt_patrolling

#include "behaviortree_cpp_v3/bt_factory.h"
BT_REGISTER_NODES(factory)
{
  factory.registerNodeType<br2_bt_patrolling::Patrol>("Patrol");
}
```

br2_bt_patrolling/src/br2_bt_patrolling/Recharge.cpp

```
#include <string>
#include <iostream>
#include <set>

#include "br2_bt_patrolling/Recharge.hpp"

#include "behaviortree_cpp_v3/behavior_tree.h"

namespace br2_bt_patrolling
{

Recharge::Recharge(
  const std::string & xml_tag_name,
  const BT::NodeConfiguration & conf)
: BT::ActionNodeBase(xml_tag_name, conf), counter_(0)
{
}

void
Recharge::halt()
{
}

BT::NodeStatus
Recharge::tick()
{
  std::cout << "Recharge tick " << counter_ << std::endl;

  if (counter_++ < 50) {
    return BT::NodeStatus::RUNNING;
  } else {
    counter_ = 0;
    config().blackboard->set<float>("battery_level", 100.0f);
    return BT::NodeStatus::SUCCESS;
  }
}

}  // namespace br2_bt_patrolling

#include "behaviortree_cpp_v3/bt_factory.h"
BT_REGISTER_NODES(factory)
{
  factory.registerNodeType<br2_bt_patrolling::Recharge>("Recharge");
}
```

br2_bt_patrolling/src/br2_bt_patrolling/GetWaypoint.cpp

```
#include <string>
#include <iostream>
#include <vector>

#include "br2_bt_patrolling/GetWaypoint.hpp"

#include "behaviortree_cpp_v3/behavior_tree.h"

#include "geometry_msgs/msg/pose_stamped.hpp"

#include "rclcpp/rclcpp.hpp"

namespace br2_bt_patrolling
{

int GetWaypoint::current_ = 0;

GetWaypoint::GetWaypoint(
  const std::string & xml_tag_name,
  const BT::NodeConfiguration & conf)
: BT::ActionNodeBase(xml_tag_name, conf)
{
  rclcpp::Node::SharedPtr node;
  config().blackboard->get("node", node);

  geometry_msgs::msg::PoseStamped wp;
  wp.header.frame_id = "map";
  wp.pose.orientation.w = 1.0;

  // recharge wp
  wp.pose.position.x = 3.67;
  wp.pose.position.y = -0.24;
  recharge_point_ = wp;
```

br2_bt_patrolling/src/br2_bt_patrolling/GetWaypoint.cpp

```
  // wp1
  wp.pose.position.x = 1.07;
  wp.pose.position.y = -12.38;
  waypoints_.push_back(wp);

  // wp2
  wp.pose.position.x = -5.32;
  wp.pose.position.y = -8.85;
  waypoints_.push_back(wp);

  // wp3
  wp.pose.position.x = -0.56;
  wp.pose.position.y = 0.24;
  waypoints_.push_back(wp);
}

void
GetWaypoint::halt()
{
}

BT::NodeStatus
GetWaypoint::tick()
{
  std::string id;
  getInput("wp_id", id);

  if (id == "recharge") {
    setOutput("waypoint", recharge_point_);
  } else {
    setOutput("waypoint", waypoints_[current_++]);
    current_ = current_ % waypoints_.size();
  }

  return BT::NodeStatus::SUCCESS;
}

}  // namespace br2_bt_patrolling

#include "behaviortree_cpp_v3/bt_factory.h"
BT_REGISTER_NODES(factory)
{
  factory.registerNodeType<br2_bt_patrolling::GetWaypoint>("GetWaypoint");
}
```

br2_bt_patrolling/src/br2_bt_patrolling/Move.cpp

```
#include <string>
#include <iostream>
#include <vector>
#include <memory>

#include "br2_bt_patrolling/Move.hpp"

#include "geometry_msgs/msg/pose_stamped.hpp"
#include "nav2_msgs/action/navigate_to_pose.hpp"

#include "behaviortree_cpp_v3/behavior_tree.h"

namespace br2_bt_patrolling
{

Move::Move(
  const std::string & xml_tag_name,
  const std::string & action_name,
  const BT::NodeConfiguration & conf)
: br2_bt_patrolling::BtActionNode<nav2_msgs::action::NavigateToPose>(xml_tag_name,
   action_name, conf)
{
}

void
Move::on_tick()
{
  geometry_msgs::msg::PoseStamped goal;
  getInput("goal", goal);

  goal_.pose = goal;
}

BT::NodeStatus
Move::on_success()
{
  RCLCPP_INFO(node_->get_logger(), "navigation Suceeded");

  return BT::NodeStatus::SUCCESS;
}


}  // namespace br2_bt_patrolling

#include "behaviortree_cpp_v3/bt_factory.h"
BT_REGISTER_NODES(factory)
{
  BT::NodeBuilder builder =
    [](const std::string & name, const BT::NodeConfiguration & config)
    {
      return std::make_unique<br2_bt_patrolling::Move>(
        name, "navigate_to_pose", config);
    };

  factory.registerBuilder<br2_bt_patrolling::Move>(
    "Move", builder);
}
```

br2_bt_patrolling/src/br2_bt_patrolling/BatteryChecker.cpp

```
#include <string>
#include <iostream>
#include <algorithm>

#include "br2_bt_patrolling/BatteryChecker.hpp"

#include "behaviortree_cpp_v3/behavior_tree.h"

#include "geometry_msgs/msg/twist.hpp"

#include "rclcpp/rclcpp.hpp"

namespace br2_bt_patrolling
{

using namespace std::chrono_literals;
using namespace std::placeholders;

BatteryChecker::BatteryChecker(
  const std::string & xml_tag_name,
  const BT::NodeConfiguration & conf)
: BT::ConditionNode(xml_tag_name, conf)
{
  config().blackboard->get("node", node_);

  vel_sub_ = node_->create_subscription<geometry_msgs::msg::Twist>(
    "/output_vel", 100, std::bind(&BatteryChecker::vel_callback, this, _1));

  last_reading_time_ = node_->now();
}

void
BatteryChecker::vel_callback(const geometry_msgs::msg::Twist::SharedPtr msg)
{
  last_twist_ = *msg;
}

void
BatteryChecker::update_battery()
{
  float battery_level;
  if (!config().blackboard->get("battery_level", battery_level)) {
    battery_level = 100.0f;
  }

  float dt = (node_->now() - last_reading_time_).seconds();
  last_reading_time_ = node_->now();

  float vel = sqrt(last_twist_.linear.x * last_twist_.linear.x +
    last_twist_.angular.z * last_twist_.angular.z);
  battery_level = std::max(0.0f, battery_level -(vel * dt * DECAY_LEVEL) - EPSILON * dt);

  config().blackboard->set("battery_level", battery_level);
}

BT::NodeStatus
BatteryChecker::tick()
{
  update_battery();

  float battery_level;
  config().blackboard->get("battery_level", battery_level);

  std::cout << battery_level << std::endl;

  if (battery_level < MIN_LEVEL) {
    return BT::NodeStatus::FAILURE;
  } else {
    return BT::NodeStatus::SUCCESS;
  }
}

}  // namespace br2_bt_patrolling

#include "behaviortree_cpp_v3/bt_factory.h"
BT_REGISTER_NODES(factory)
{
  factory.registerNodeType<br2_bt_patrolling::BatteryChecker>("BatteryChecker");
}
```

br2_bt_patrolling/src/br2_bt_patrolling/TrackObjects.cpp

```
#include <string>
#include <iostream>
#include <vector>
#include <memory>

#include "br2_bt_patrolling/TrackObjects.hpp"

#include "geometry_msgs/msg/pose_stamped.hpp"
#include "nav2_msgs/action/navigate_to_pose.hpp"

#include "behaviortree_cpp_v3/behavior_tree.h"

namespace br2_bt_patrolling
{

TrackObjects::TrackObjects(
  const std::string & xml_tag_name,
  const std::string & action_name,
  const BT::NodeConfiguration & conf)
: br2_bt_patrolling::BtLifecycleCtrlNode(xml_tag_name, action_name, conf)
{
}

}  // namespace br2_bt_patrolling

#include "behaviortree_cpp_v3/bt_factory.h"
BT_REGISTER_NODES(factory)
{
  BT::NodeBuilder builder =
    [](const std::string & name, const BT::NodeConfiguration & config)
    {
      return std::make_unique<br2_bt_patrolling::TrackObjects>(
        name, "/head_tracker", config);
    };

  factory.registerBuilder<br2_bt_patrolling::TrackObjects>(
    "TrackObjects", builder);
}
```

br2_bt_patrolling/src/patrolling_main.cpp

```
#include <string>
#include <memory>

#include "behaviortree_cpp_v3/behavior_tree.h"
#include "behaviortree_cpp_v3/bt_factory.h"
#include "behaviortree_cpp_v3/utils/shared_library.h"
#include "behaviortree_cpp_v3/loggers/bt_zmq_publisher.h"

#include "ament_index_cpp/get_package_share_directory.hpp"

#include "rclcpp/rclcpp.hpp"


int main(int argc, char * argv[])
{
  rclcpp::init(argc, argv);

  auto node = rclcpp::Node::make_shared("patrolling_node");

  BT::BehaviorTreeFactory factory;
  BT::SharedLibrary loader;

  factory.registerFromPlugin(loader.getOSName("br2_battery_checker_bt_node"));
  factory.registerFromPlugin(loader.getOSName("br2_patrol_bt_node"));
  factory.registerFromPlugin(loader.getOSName("br2_recharge_bt_node"));
  factory.registerFromPlugin(loader.getOSName("br2_move_bt_node"));
  factory.registerFromPlugin(loader.getOSName("br2_get_waypoint_bt_node"));
  factory.registerFromPlugin(loader.getOSName("br2_track_objects_bt_node"));

  std::string pkgpath = ament_index_cpp::get_package_share_directory("br2_bt_patrolling");
  std::string xml_file = pkgpath + "/behavior_tree_xml/patrolling.xml";

  auto blackboard = BT::Blackboard::create();
  blackboard->set("node", node);
  BT::Tree tree = factory.createTreeFromFile(xml_file, blackboard);

  auto publisher_zmq = std::make_shared<BT::PublisherZMQ>(tree, 10, 2666, 2667);

  rclcpp::Rate rate(10);

  bool finish = false;
  while (!finish && rclcpp::ok()) {
    finish = tree.rootNode()->executeTick() == BT::NodeStatus::SUCCESS;

    rclcpp::spin_some(node);
    rate.sleep();
  }

  rclcpp::shutdown();
  return 0;
}
```

br2_bt_patrolling/tests/bt_action_test.cpp

```
#include <string>
#include <list>
#include <memory>
#include <vector>
#include <set>

#include "behaviortree_cpp_v3/behavior_tree.h"
#include "behaviortree_cpp_v3/bt_factory.h"
#include "behaviortree_cpp_v3/utils/shared_library.h"

#include "ament_index_cpp/get_package_share_directory.hpp"

#include "geometry_msgs/msg/twist.hpp"
#include "nav2_msgs/action/navigate_to_pose.hpp"
#include "lifecycle_msgs/msg/transition.hpp"
#include "lifecycle_msgs/msg/state.hpp"

#include "rclcpp/rclcpp.hpp"
#include "rclcpp_lifecycle/lifecycle_node.hpp"
#include "rclcpp_action/rclcpp_action.hpp"

#include "br2_bt_patrolling/TrackObjects.hpp"

#include "gtest/gtest.h"

using namespace std::placeholders;
using namespace std::chrono_literals;


class VelocitySinkNode : public rclcpp::Node
{
public:
  VelocitySinkNode()
  : Node("VelocitySink")
  {
    vel_sub_ = create_subscription<geometry_msgs::msg::Twist>(
      "/output_vel", 100, std::bind(&VelocitySinkNode::vel_callback, this, _1));
  }

  void vel_callback(geometry_msgs::msg::Twist::SharedPtr msg)
  {
    vel_msgs_.push_back(*msg);
  }

  std::list<geometry_msgs::msg::Twist> vel_msgs_;

private:
  rclcpp::Subscription<geometry_msgs::msg::Twist>::SharedPtr vel_sub_;
};

class Nav2FakeServer : public rclcpp::Node
{
  using NavigateToPose = nav2_msgs::action::NavigateToPose;
  using GoalHandleNavigateToPose = rclcpp_action::ServerGoalHandle<NavigateToPose>;

public:
  Nav2FakeServer()
  : Node("nav2_fake_server_node") {}

  void start_server()
  {
    move_action_server_ = rclcpp_action::create_server<NavigateToPose>(
      shared_from_this(),
      "navigate_to_pose",
      std::bind(&Nav2FakeServer::handle_goal, this, _1, _2),
      std::bind(&Nav2FakeServer::handle_cancel, this, _1),
      std::bind(&Nav2FakeServer::handle_accepted, this, _1));
  }

private:
  rclcpp_action::Server<NavigateToPose>::SharedPtr move_action_server_;

  rclcpp_action::GoalResponse handle_goal(
    const rclcpp_action::GoalUUID & uuid,
    std::shared_ptr<const NavigateToPose::Goal> goal)
  {
    return rclcpp_action::GoalResponse::ACCEPT_AND_EXECUTE;
  }
```

br2_bt_patrolling/tests/bt_action_test.cpp

```
  rclcpp_action::CancelResponse handle_cancel(
    const std::shared_ptr<GoalHandleNavigateToPose> goal_handle)
  {
    return rclcpp_action::CancelResponse::ACCEPT;
  }

  void handle_accepted(const std::shared_ptr<GoalHandleNavigateToPose> goal_handle)
  {
    std::thread{std::bind(&Nav2FakeServer::execute, this, _1), goal_handle}.detach();
  }

  void execute(const std::shared_ptr<GoalHandleNavigateToPose> goal_handle)
  {
    auto feedback = std::make_shared<NavigateToPose::Feedback>();
    auto result = std::make_shared<NavigateToPose::Result>();

    auto start = now();

    while ((now() - start) < 5s) {
      feedback->distance_remaining = 5.0 - (now() - start).seconds();
      goal_handle->publish_feedback(feedback);
    }

    goal_handle->succeed(result);
  }
};

class StoreWP : public BT::ActionNodeBase
{
public:
  explicit StoreWP(
    const std::string & xml_tag_name,
    const BT::NodeConfiguration & conf)
  : BT::ActionNodeBase(xml_tag_name, conf) {}

  void halt() {}
  BT::NodeStatus tick()
  {
    waypoints_.push_back(getInput<geometry_msgs::msg::PoseStamped>("in").value());
    return BT::NodeStatus::SUCCESS;
  }

  static BT::PortsList providedPorts()
  {
    return BT::PortsList(
    {
      BT::InputPort<geometry_msgs::msg::PoseStamped>("in")
    });
  }

  static std::vector<geometry_msgs::msg::PoseStamped> waypoints_;
};

std::vector<geometry_msgs::msg::PoseStamped> StoreWP::waypoints_;

TEST(bt_action, recharge_btn)
{
  auto node = rclcpp::Node::make_shared("recharge_btn_node");

  BT::BehaviorTreeFactory factory;
  BT::SharedLibrary loader;

  factory.registerFromPlugin(loader.getOSName("br2_recharge_bt_node"));

  std::string xml_bt =
    R"(
    <root main_tree_to_execute = "MainTree" >
      <BehaviorTree ID="MainTree">
          <Recharge    name="recharge"/>
      </BehaviorTree>
    </root>)";

  auto blackboard = BT::Blackboard::create();
  blackboard->set("node", node);
  BT::Tree tree = factory.createTreeFromText(xml_bt, blackboard);

  rclcpp::Rate rate(10);
```

br2_bt_patrolling/tests/bt_action_test.cpp

```
  bool finish = false;
  while (!finish && rclcpp::ok()) {
    finish = tree.rootNode()->executeTick() == BT::NodeStatus::SUCCESS;
    rate.sleep();
  }

  float battery_level;
  ASSERT_TRUE(blackboard->get("battery_level", battery_level));
  ASSERT_NEAR(battery_level, 100.0f, 0.0000001);
}

TEST(bt_action, patrol_btn)
{
  auto node = rclcpp::Node::make_shared("patrol_btn_node");
  auto node_sink = std::make_shared<VelocitySinkNode>();

  BT::BehaviorTreeFactory factory;
  BT::SharedLibrary loader;

  factory.registerFromPlugin(loader.getOSName("br2_patrol_bt_node"));

  std::string xml_bt =
    R"(
    <root main_tree_to_execute = "MainTree" >
      <BehaviorTree ID="MainTree">
          <Patrol    name="patrol"/>
      </BehaviorTree>
    </root>)";

  auto blackboard = BT::Blackboard::create();
  blackboard->set("node", node);
  BT::Tree tree = factory.createTreeFromText(xml_bt, blackboard);

  rclcpp::Rate rate(10);

  bool finish = false;
  int counter = 0;
  while (!finish && rclcpp::ok()) {
    finish = tree.rootNode()->executeTick() == BT::NodeStatus::SUCCESS;
    rclcpp::spin_some(node_sink->get_node_base_interface());
    rate.sleep();
  }

  ASSERT_FALSE(node_sink->vel_msgs_.empty());
  ASSERT_NEAR(node_sink->vel_msgs_.size(), 150, 2);

  geometry_msgs::msg::Twist & one_twist = node_sink->vel_msgs_.front();

  ASSERT_GT(one_twist.angular.z, 0.1);
  ASSERT_NEAR(one_twist.linear.x, 0.0, 0.0000001);
}

TEST(bt_action, move_btn)
{
  auto node = rclcpp::Node::make_shared("move_btn_node");
  auto nav2_fake_node = std::make_shared<Nav2FakeServer>();

  nav2_fake_node->start_server();

  bool finish = false;
  std::thread t([&]() {
      while (!finish) {rclcpp::spin_some(nav2_fake_node);}
    });


  BT::BehaviorTreeFactory factory;
  BT::SharedLibrary loader;

  factory.registerFromPlugin(loader.getOSName("br2_move_bt_node"));
```

br2_bt_patrolling/tests/bt_action_test.cpp

```
  std::string xml_bt =
    R"(
    <root main_tree_to_execute = "MainTree" >
      <BehaviorTree ID="MainTree">
          <Move    name="move" goal="{goal}"/>
      </BehaviorTree>
    </root>)";

  auto blackboard = BT::Blackboard::create();
  blackboard->set("node", node);

  geometry_msgs::msg::PoseStamped goal;
  blackboard->set("goal", goal);

  BT::Tree tree = factory.createTreeFromText(xml_bt, blackboard);

  rclcpp::Rate rate(10);

  int counter = 0;
  while (!finish && rclcpp::ok()) {
    finish = tree.rootNode()->executeTick() == BT::NodeStatus::SUCCESS;
    rate.sleep();
  }

  t.join();
}

TEST(bt_action, get_waypoint_btn)
{
  auto node = rclcpp::Node::make_shared("get_waypoint_btn_node");

  rclcpp::spin_some(node);

  {
    BT::BehaviorTreeFactory factory;
    BT::SharedLibrary loader;

    factory.registerFromPlugin(loader.getOSName("br2_get_waypoint_bt_node"));

    std::string xml_bt =
      R"(
      <root main_tree_to_execute = "MainTree" >
        <BehaviorTree ID="MainTree">
          <GetWaypoint    name="recharge" wp_id="{id}" waypoint="{waypoint}"/>
        </BehaviorTree>
      </root>)";

    auto blackboard = BT::Blackboard::create();
    blackboard->set("node", node);
    blackboard->set<std::string>("id", "recharge");

    BT::Tree tree = factory.createTreeFromText(xml_bt, blackboard);

    rclcpp::Rate rate(10);

    bool finish = false;
    int counter = 0;
    while (!finish && rclcpp::ok()) {
      finish = tree.rootNode()->executeTick() == BT::NodeStatus::SUCCESS;
      counter++;
      rate.sleep();
    }

    auto point = blackboard->get<geometry_msgs::msg::PoseStamped>("waypoint");

    ASSERT_EQ(counter, 1);
    ASSERT_NEAR(point.pose.position.x, 3.67, 0.0000001);
    ASSERT_NEAR(point.pose.position.y, -0.24, 0.0000001);
  }

  {
    BT::BehaviorTreeFactory factory;
    BT::SharedLibrary loader;

    factory.registerNodeType<StoreWP>("StoreWP");
    factory.registerFromPlugin(loader.getOSName("br2_get_waypoint_bt_node"));
```

br2_bt_patrolling/tests/bt_action_test.cpp

```
    std::string xml_bt =
      R"(
      <root main_tree_to_execute = "MainTree" >
        <BehaviorTree ID="MainTree">
          <Sequence name="root_sequence">
             <GetWaypoint    name="wp1" wp_id="next" waypoint="{waypoint}"/>
             <StoreWP in="{waypoint}"/>
             <GetWaypoint    name="wp2" wp_id="next" waypoint="{waypoint}"/>
             <StoreWP in="{waypoint}"/>
             <GetWaypoint    name="wp3" wp_id="" waypoint="{waypoint}"/>
             <StoreWP in="{waypoint}"/>
             <GetWaypoint    name="wp4" wp_id="recharge" waypoint="{waypoint}"/>
             <StoreWP in="{waypoint}"/>
             <GetWaypoint    name="wp5" wp_id="wp1" waypoint="{waypoint}"/>
             <StoreWP in="{waypoint}"/>
             <GetWaypoint    name="wp6" wp_id="wp2" waypoint="{waypoint}"/>
             <StoreWP in="{waypoint}"/>
             <GetWaypoint    name="wpt" waypoint="{waypoint}"/>
             <StoreWP in="{waypoint}"/>
          </Sequence>
        </BehaviorTree>
      </root>)";

    auto blackboard = BT::Blackboard::create();
    blackboard->set("node", node);

    BT::Tree tree = factory.createTreeFromText(xml_bt, blackboard);

    rclcpp::Rate rate(10);

    bool finish = false;
    while (!finish && rclcpp::ok()) {
      finish = tree.rootNode()->executeTick() == BT::NodeStatus::SUCCESS;
      rate.sleep();
    }

    const auto & waypoints = StoreWP::waypoints_;
    ASSERT_EQ(waypoints.size(), 7);
    ASSERT_NEAR(waypoints[0].pose.position.x, 1.07, 0.0000001);
    ASSERT_NEAR(waypoints[0].pose.position.y, -12.38, 0.0000001);
    ASSERT_NEAR(waypoints[1].pose.position.x, -5.32, 0.0000001);
    ASSERT_NEAR(waypoints[1].pose.position.y, -8.85, 0.0000001);
    ASSERT_NEAR(waypoints[2].pose.position.x, -0.56, 0.0000001);
    ASSERT_NEAR(waypoints[2].pose.position.y, 0.24, 0.0000001);

    ASSERT_NEAR(waypoints[3].pose.position.x, 3.67, 0.0000001);
    ASSERT_NEAR(waypoints[3].pose.position.y, -0.24, 0.0000001);

    ASSERT_NEAR(waypoints[4].pose.position.x, 1.07, 0.0000001);
    ASSERT_NEAR(waypoints[4].pose.position.y, -12.38, 0.0000001);
    ASSERT_NEAR(waypoints[5].pose.position.x, -5.32, 0.0000001);
    ASSERT_NEAR(waypoints[5].pose.position.y, -8.85, 0.0000001);
    ASSERT_NEAR(waypoints[6].pose.position.x, -0.56, 0.0000001);
    ASSERT_NEAR(waypoints[6].pose.position.y, 0.24, 0.0000001);
  }
}

TEST(bt_action, battery_checker_btn)
{
  auto node = rclcpp::Node::make_shared("battery_checker_btn_node");
  auto vel_pub = node->create_publisher<geometry_msgs::msg::Twist>("/output_vel", 100);

  BT::BehaviorTreeFactory factory;
  BT::SharedLibrary loader;

  factory.registerFromPlugin(loader.getOSName("br2_battery_checker_bt_node"));
  factory.registerFromPlugin(loader.getOSName("br2_patrol_bt_node"));

  std::string xml_bt =
    R"(
    <root main_tree_to_execute = "MainTree" >
      <BehaviorTree ID="MainTree">
          <ReactiveSequence>
              <BatteryChecker    name="battery_checker"/>
              <Patrol    name="patrol"/>
          </ReactiveSequence>
      </BehaviorTree>
    </root>)";
```

br2_bt_patrolling/tests/bt_action_test.cpp

```
  auto blackboard = BT::Blackboard::create();
  blackboard->set("node", node);
  BT::Tree tree = factory.createTreeFromText(xml_bt, blackboard);

  rclcpp::Rate rate(10);
  geometry_msgs::msg::Twist vel;
  vel.linear.x = 0.8;

  bool finish = false;
  int counter = 0;
  while (!finish && rclcpp::ok()) {
    finish = tree.rootNode()->executeTick() == BT::NodeStatus::SUCCESS;

    vel_pub->publish(vel);

    rclcpp::spin_some(node);
    rate.sleep();
  }

  float battery_level;
  ASSERT_TRUE(blackboard->get("battery_level", battery_level));
  ASSERT_NEAR(battery_level, 94.6, 1.0);
}

TEST(bt_action, track_objects_btn_1)
{
  auto node = rclcpp::Node::make_shared("track_objects_btn_node");
  auto node_head_tracker = rclcpp_lifecycle::LifecycleNode::make_shared("head_tracker");

  bool finish = false;
  std::thread t([&]() {
      while (!finish) {rclcpp::spin_some(node_head_tracker->get_node_base_interface());}
    });

  BT::NodeConfiguration conf;
  conf.blackboard = BT::Blackboard::create();
  conf.blackboard->set("node", node);
  br2_bt_patrolling::BtLifecycleCtrlNode bt_node("TrackObjects", "head_tracker", conf);

  bt_node.change_state_client_ = bt_node.createServiceClient<lifecycle_msgs::srv::
   ChangeState>(
    "/head_tracker/change_state");
  ASSERT_TRUE(bt_node.change_state_client_->service_is_ready());

  bt_node.get_state_client_ = bt_node.createServiceClient<lifecycle_msgs::srv::GetState>(
    "/head_tracker/get_state");
  ASSERT_TRUE(bt_node.get_state_client_->service_is_ready());
  auto start = node->now();

  rclcpp::Rate rate(10);
  while (rclcpp::ok() && (node->now() - start) < 1s) {
    rclcpp::spin_some(node);
    rate.sleep();
  }

  ASSERT_EQ(bt_node.get_state(), lifecycle_msgs::msg::State::PRIMARY_STATE_UNCONFIGURED);
  bt_node.ctrl_node_state_ = lifecycle_msgs::msg::State::PRIMARY_STATE_UNCONFIGURED;
  ASSERT_FALSE(bt_node.set_state(lifecycle_msgs::msg::State::PRIMARY_STATE_ACTIVE));

  node_head_tracker->trigger_transition(lifecycle_msgs::msg::Transition::
   TRANSITION_CONFIGURE);

  start = node->now();
  while (rclcpp::ok() && (node->now() - start) < 1s) {
    rclcpp::spin_some(node);
    rate.sleep();
  }

  bt_node.ctrl_node_state_ = bt_node.get_state();

  ASSERT_TRUE(bt_node.set_state(lifecycle_msgs::msg::State::PRIMARY_STATE_ACTIVE));
  ASSERT_EQ(bt_node.get_state(), lifecycle_msgs::msg::State::PRIMARY_STATE_ACTIVE);

  start = node->now();
  while (rclcpp::ok() && (node->now() - start) < 1s) {
    rclcpp::spin_some(node);
    rate.sleep();
  }
```

br2_bt_patrolling/tests/bt_action_test.cpp

```
  bt_node.ctrl_node_state_ = bt_node.get_state();

  ASSERT_TRUE(bt_node.set_state(lifecycle_msgs::msg::State::PRIMARY_STATE_INACTIVE));
  ASSERT_EQ(bt_node.get_state(), lifecycle_msgs::msg::State::PRIMARY_STATE_INACTIVE);

  finish = true;
  t.join();
}

TEST(bt_action, track_objects_btn_2)
{
  auto node = rclcpp::Node::make_shared("track_objects_btn_node");
  auto node_head_tracker = rclcpp_lifecycle::LifecycleNode::make_shared("head_tracker");

  bool finish = false;
  std::thread t([&]() {
      while (!finish) {rclcpp::spin_some(node_head_tracker->get_node_base_interface());}
    });

  BT::NodeConfiguration conf;
  conf.blackboard = BT::Blackboard::create();
  conf.blackboard->set("node", node);
  br2_bt_patrolling::BtLifecycleCtrlNode bt_node("TrackObjects", "head_tracker", conf);

  node_head_tracker->trigger_transition(lifecycle_msgs::msg::Transition::
   TRANSITION_CONFIGURE);

  rclcpp::Rate rate(10);
  auto start = node->now();
  while (rclcpp::ok() && (node->now() - start) < 1s) {
    rclcpp::spin_some(node);
    rate.sleep();
  }

  ASSERT_EQ(bt_node.tick(), BT::NodeStatus::RUNNING);

  ASSERT_TRUE(bt_node.change_state_client_->service_is_ready());
  ASSERT_TRUE(bt_node.get_state_client_->service_is_ready());

  ASSERT_EQ(bt_node.get_state(), lifecycle_msgs::msg::State::PRIMARY_STATE_ACTIVE);

  ASSERT_EQ(bt_node.tick(), BT::NodeStatus::RUNNING);

  bt_node.halt();

  start = node->now();
  while (rclcpp::ok() && (node->now() - start) < 1s) {
    rclcpp::spin_some(node);
    rate.sleep();
  }

  ASSERT_EQ(bt_node.get_state(), lifecycle_msgs::msg::State::PRIMARY_STATE_INACTIVE);

  finish = true;
  t.join();
}

TEST(bt_action, track_objects_btn_3)
{
  auto node = rclcpp::Node::make_shared("track_objects_btn_node");
  auto node_head_tracker = rclcpp_lifecycle::LifecycleNode::make_shared("head_tracker");

  node_head_tracker->trigger_transition(lifecycle_msgs::msg::Transition::
   TRANSITION_CONFIGURE);

  bool finish = false;
  std::thread t([&]() {
      while (!finish) {rclcpp::spin_some(node_head_tracker->get_node_base_interface());}
    });

  BT::BehaviorTreeFactory factory;
  BT::SharedLibrary loader;

  factory.registerFromPlugin(loader.getOSName("br2_track_objects_bt_node"));
```

br2_bt_patrolling/tests/bt_action_test.cpp

```
  std::string xml_bt =
    R"(
    <root main_tree_to_execute = "MainTree" >
      <BehaviorTree ID="MainTree">
          <KeepRunningUntilFailure>
              <TrackObjects    name="track_objects"/>
          </KeepRunningUntilFailure>
      </BehaviorTree>
    </root>)";

  auto blackboard = BT::Blackboard::create();
  blackboard->set("node", node);
  auto start = node->now();
  rclcpp::Rate rate(10);

  {
    BT::Tree tree = factory.createTreeFromText(xml_bt, blackboard);

    ASSERT_EQ(
      node_head_tracker->get_current_state().id(),
      lifecycle_msgs::msg::State::PRIMARY_STATE_INACTIVE);

    while (rclcpp::ok() && (node->now() - start) < 1s) {
      tree.rootNode()->executeTick() == BT::NodeStatus::RUNNING;

      rclcpp::spin_some(node);
      rate.sleep();
    }
    ASSERT_EQ(
      node_head_tracker->get_current_state().id(),
      lifecycle_msgs::msg::State::PRIMARY_STATE_ACTIVE);
  }

  start = node->now();
  while (rclcpp::ok() && (node->now() - start) < 1s) {
    rclcpp::spin_some(node);
    rate.sleep();
  }

  ASSERT_EQ(
    node_head_tracker->get_current_state().id(),
    lifecycle_msgs::msg::State::PRIMARY_STATE_INACTIVE);

  finish = true;
  t.join();
}

TEST(bt_action, move_track_btn)
{
  auto node = rclcpp::Node::make_shared("move_btn_node");
  auto nav2_fake_node = std::make_shared<Nav2FakeServer>();
  auto node_head_tracker = rclcpp_lifecycle::LifecycleNode::make_shared("head_tracker");

  node_head_tracker->trigger_transition(lifecycle_msgs::msg::Transition::
   TRANSITION_CONFIGURE);

  nav2_fake_node->start_server();

  rclcpp::executors::SingleThreadedExecutor exe;
  exe.add_node(nav2_fake_node);
  exe.add_node(node_head_tracker->get_node_base_interface());
  bool finish = false;
  std::thread t([&]() {
      while (!finish) {exe.spin_some();}
    });

  BT::BehaviorTreeFactory factory;
  BT::SharedLibrary loader;

  factory.registerFromPlugin(loader.getOSName("br2_move_bt_node"));
  factory.registerFromPlugin(loader.getOSName("br2_track_objects_bt_node"));
```

br2_bt_patrolling/tests/bt_action_test.cpp

```
  std::string xml_bt =
    R"(
    <root main_tree_to_execute = "MainTree" >
      <BehaviorTree ID="MainTree">
          <Parallel success_threshold="1" failure_threshold="1">
            <TrackObjects    name="track_objects"/>
            <Move    name="move" goal="{goal}"/>
          </Parallel>
      </BehaviorTree>
    </root>)";

  auto blackboard = BT::Blackboard::create();
  blackboard->set("node", node);

  geometry_msgs::msg::PoseStamped goal;
  blackboard->set("goal", goal);

  BT::Tree tree = factory.createTreeFromText(xml_bt, blackboard);

  ASSERT_EQ(
    node_head_tracker->get_current_state().id(),
    lifecycle_msgs::msg::State::PRIMARY_STATE_INACTIVE);

  rclcpp::Rate rate(10);
  auto start = node->now();
  auto finish_tree = false;
  while (rclcpp::ok() && (node->now() - start) < 1s) {
    finish_tree = tree.rootNode()->executeTick() == BT::NodeStatus::SUCCESS;

    rclcpp::spin_some(node);
    rate.sleep();
  }

  ASSERT_FALSE(finish_tree);
  ASSERT_EQ(
    node_head_tracker->get_current_state().id(),
    lifecycle_msgs::msg::State::PRIMARY_STATE_ACTIVE);

  while (rclcpp::ok() && !finish_tree) {
    finish_tree = tree.rootNode()->executeTick() == BT::NodeStatus::SUCCESS;

    rclcpp::spin_some(node);
    rate.sleep();
  }

  start = node->now();
  while (rclcpp::ok() && (node->now() - start) < 1s) {
    rclcpp::spin_some(node);
    rate.sleep();
  }

  ASSERT_EQ(
    node_head_tracker->get_current_state().id(),
    lifecycle_msgs::msg::State::PRIMARY_STATE_INACTIVE);

  finish = true;
  t.join();
}

int main(int argc, char ** argv)
{
  rclcpp::init(argc, argv);

  testing::InitGoogleTest(&argc, argv);
  return RUN_ALL_TESTS();
}
```

br2_bt_patrolling/tests/CMakeLists.txt

```
ament_add_gtest(bt_action_test bt_action_test.cpp)
ament_target_dependencies(bt_action_test ${dependencies})
target_link_libraries(bt_action_test br2_track_objects_bt_node)
```

Bibliography

[1] Rodney A. Brooks. Elephants don't play chess. *Robotics and Autonomous Systems*, 6(1):3–15, 1990. Designing Autonomous Agents.

[2] Brian Gerkey, Richard T Vaughan, Andrew Howard, et al. The player/stage project: Tools for multi-robot and distributed sensor systems. In *Proceedings of the 11th international conference on advanced robotics*, volume 1, pages 317–323. Citeseer, 2003.

[3] Steven Macenski, Francisco Martin, Ruffin White, and Jonatan Ginés Clavero. The marathon 2: A navigation system. In *2020 IEEE/RSJ International Conference on Intelligent Robots and Systems (IROS)*, 2020.

[4] Alejandro Marzinotto, Michele Colledanchise, Christian Smith, and Petter Ögren. Towards a unified behavior trees framework for robot control. In *2014 IEEE International Conference on Robotics and Automation (ICRA)*, pages 5420–5427, 2014.

[5] Giorgio Metta, Paul Fitzpatrick, and Lorenzo Natale. Yarp: Yet another robot platform. *International Journal of Advanced Robotic Systems*, 3(1):8, 2006.

[6] Michael Montemerlo, Nicholas Roy, and Sebastian Thrun. Perspectives on standardization in mobile robot programming: The carnegie mellon navigation (carmen) toolkit. In *Proceedings 2003 IEEE/RSJ International Conference on Intelligent Robots and Systems (IROS 2003)(Cat. No. 03CH37453)*, volume 3, pages 2436–2441. IEEE, 2003.

[7] Morgan Quigley, Brian Gerkey, Ken Conley, Josh Faust, Tully Foote, Jeremy Leibs, Eric Berger, Rob Wheeler, and Andrew Ng. Ros: an open-source robot operating system. In *Proc. of the IEEE Intl. Conf. on Robotics and Automation (ICRA) Workshop on Open Source Robotics*, Kobe, Japan, May 2009.

[8] Dirk Thomas, William Woodall, and Esteve Fernandez. Next-generation ros: Building on dds. In OSRF, editor, *ROSCon 2014*, 2014.

[9] Sebastian Thrun, Dieter Fox, Wolfram Burgard, and Frank Dellaert. Robust monte carlo localization for mobile robots. *Artificial Intelligence*, 128(1):99–141, 2001.

Index

9781032267203